Fossils

Fossils

300 of the Earth's fossilized species

General editor
Carl Mehling

Fossils

300 of the Earth's fossilized species

General Editor:
Carl Mehling

THUNDER BAY
P · R · E · S · S

San Diego, California

Thunder Bay Press
An imprint of the Advantage Publishers Group
5880 Oberlin Drive, San Diego, CA 92121-4794
www.thunderbaybooks.com

Amber Books Ltd
Bradley's Close
74–77 White Lion Street
London N1 9PF
www.amberbooks.co.uk

ISBN-13: 978-1-59223-737-1
ISBN-10: 1-59223-737-1

Library of Congress Cataloging-in-Publication Data available upon request.

Printed in Singapore
1 2 3 4 5 11 10 09 08 07

Project Editor: Sarah Uttridge
Design: Joe Conneally

PICTURE CREDITS

All pictures © DeAgostini Picture Library except the following: Mick Ellison: 151, 166; Mike
Everhart: 83, 226; Carl Mehling: 24, 26, 27, 28, 29, 35, 37, 38, 40, 41, 42, 43, 46, 47, 52, 77, 87,
88, 96, 115, 119, 131, 155, 157, 158, 159, 160, 167, 168, 180, 181, 185, 221, 232, 233, 235,
236, 238, 239, 245, 246, 251, 252, 260, 262, 265, 267, 271, 272, 273, 274, 276, 278, 279, 282,
283, 284, 286, 289, 303, 304, 305, 307, 308, 309.

CONTENTS

Introduction

Museum mounts, like this fossil proboscidean, are an important way of reaching and inspiring the public to learn about palaeontology.

Fossils have intrigued the inquisitive for millennia. They have been found on the stone tools of some of our very early human ancestors, protected from the damage of knapping and thus clearly meant to adorn these pieces. They have been used as decoration and sacred objects in countless contexts throughout recorded history. Some have considered them as works of the Devil, created to deceive mankind, or as creations of God, meant to fascinate, confuse or even test faith. They have even been used for magic and medicine. Today, fossils are generally viewed as time travellers crossing the abyss of time with stories to tell from the other side; humbling stories that remind us of our place in an unfathomably long history of life on this planet – continuing stories.

But what exactly is a fossil? Just like the organisms they now clearly represent, the idea of 'fossil' has evolved over time to adapt to the environment – in this case, the environment of thought and theory. The word fossil is derived from Latin and refers to things 'dug up'. Originally, this could refer to the remains of past life, artefacts or even crystals. Currently, the definition has become restricted to denote any evidence of ancient life, 'ancient' meaning 'from a past geological time period', which translates as 'older than 10,000 years old'. Many assume 'fossil' indicates 'turned to stone', but, as readers will see, there are cases where this is not so.

Many books state that fossils need to be rapidly buried, and although this is certainly true for the overwhelming majority, there are a few cases in which this need not occur, most notably desiccated Pleistocene sloths found in South American caves. But because sedimentary entombment is such a superior way to ensure long-term preservation, it should come as no surprise that marine taxa are much better represented in the fossil record than are terrestrial taxa: the sea is a legendary, constant and reliable burier of things. Readers will notice how significantly marine taxa are represented in this book. Interestingly, it should also be noted how many marine taxa are also global in distribution. This is due to the comparatively unrestricted access marine organisms have to other similar habitats, whereas terrestrial organisms are often prevented from reaching habitats similar to their own on other continents by what is to them a vast, uncrossable ocean.

SCIENCE

Since palaeontology, or the study of fossils, is a science, it is always growing and updating, improving its ideas and striving to reduce inaccuracies. As part of that continuum, the creation of this book was no exception. But books are static capsules of information, and capturing a moving, living entity like palaeontology on fixed pages is difficult. Astute readers will find the inevitable errors, especially as time passes; but this, too, is science at work and is encouraged by the editors.

To geologists, exposures such as this are open pages in Earth's history waiting to be read and understood.

INTRODUCTION

Fossils are Nature's record of the life forms that inhabited Earth in geologically ancient times.

An especially dynamic area touched on in this book is classification: the act of categorizing organisms and its resultant illumination of their evolutionary origins and relationships. This nested series of categories is best represented by a riotously branching tree. Books are, by nature, linear, and the problems with translating a tree into a linear format should be remembered as one reads the text of any book tracing this tree. Being an extremely active area of biological science, classification is particularly hard to pin down, especially since the work of so many different researchers often yields slightly differing, or even conflicting, classifications. Editorial choices were made in order to provide a starting point for readers who want to look more deeply into this subject.

Time ranges for taxa are also quite changeable. One reason is that resolution of the placement of boundaries between time periods is always being enhanced, as is the clarification of the absolute dates for these time periods. Thus, time ranges for any of the organisms herein may be extended or abridged with a deeper look at the current scientific literature as well as future fossil research. These entries are also meant as a guide for readers intrigued by the intricacies of stratigraphy and geologic time.

Palaeontology is a fusion of biology and geology, since fossils represent organisms but are found in a geological context. A thorough understanding of fossils as objects requires understanding of at least these two topics.

BIOLOGY

Organismal remains from deep time are nearly all the evidence we have left of long-extinct species. First and foremost, they answer many of our questions about anatomy, especially hard-part anatomy (bones, shells, wood, etc), which has the best potential for becoming a fossil. From this launching point, knowledge of living organisms, especially those closely related to the fossil in question, can join in and offer suggestions as to the physiology, biomechanics and behaviour of the ancient inhabitants of the earth. And when put together with deductions of other contemporary and local extinct forms – an ancient organismal community – fossils help paint pictures of palaeoecology. And since innovation and change over time are salient features of life as we currently know it, fossils have long been icons in the field of evolution. And by confirming, clarifying and often contradicting evolutionary scenarios derived from exclusively extant organisms, fossils often prompt the next moves in the science of evolution. A crucial element of evolution is extinction. Not only can the changes of organisms, and groups thereof, into new types be chronicled and documented, but their disappearances, sometime on a colossal scale, appear in the rock record and flesh out our view of the past and of recent extinctions, which reflect not only the natural species disappearances recorded throughout the planet's history, but also the accelerated rate of extinction influenced by our own species in modern times.

GEOLOGY

Fossils in most forms can also be treated as rock types or particles within the greater framework of geology. When viewed as geological entities, they can be studied to help delimit geologic time periods or elucidate palaeoenvironments and even palaeogeography. The field of plate tectonics was, in part, born from the information that fossils offered about the movement of continents.

Taphonomy (the study of the processes of fossilization) begins with the path that organisms take after death, on their way to becoming fossils. It also covers what happens to their traces after they are made by the still-living organism, like the ways footprints or coprolites are rescued from oblivion. Many of the details of taphonomy are biological, such as the decay and scavenging of dead bodies, but these also tend to have a profound effect on the way the future fossil's matrix behaves. The chemicals of decay and mineralization, and how they work, are important in understanding the origins of different fossil types and the geologic context in which they are found.

Palaeontologists piece together fragments of organisms, like these leaves, to reconstruct extinct life forms.

PERSPECTIVE

In the making of this book, special attention was also paid to providing a well-rounded account of the whole of palaeontology. Extra awareness was given to areas all too often overlooked. It is in these areas that the true hidden gems of any discipline are often found. Bizarre stories of etymology (word origins), collection, palaeopathology, history and folklore are visited and woven into the accounts of the biology of the actual organisms and the geological contexts in which they are found.

The oldest records of life on earth reach back a staggering 4.5 billion years. From its inauspicious beginnings, life has spread to occupy every conceivable climate, altitude, medium and substrate, at sizes ranging from invisibly tiny to inconceivably titanic, and often with populations numbering in the trillions of individuals. Today's organismal diversity is sometimes estimated at 100,000,000 species. And this is only a brief whisper in the discourse of billions of years of evolution. At any moment in geologic time, at least for the last 500 million years or so (the Phanerozoic), this same level of diversity is to be expected.

Because they are rare, informative and often beautiful, complete skeletons are among the most attractive fossils.

Fossils of some organisms are stupendously common – the White Cliffs of Dover are solidly composed of countless trillions of fossil microbes – but when put against the whole history of living things, the chance of any organism being preserved until the present day as a fossil is abysmally small. Any book like this can survey only minuscule subsets of this unfathomable history. Yet, from a human perspective, we have more fossils to describe, interpret and otherwise muse over than we can ever hope to finish working with.

COLLECTING AND COLLECTIONS

This brings us to another area not well covered in popular accounts of palaeontology. Fossils are of little use until found and interpreted. Despite rumours of the utility of ground-penetrating radar and other deep-imaging technologies, fossils are still almost exclusively located with the trained, or merely lucky, eyes of the fossil hunter. And it is usually necessary for fossils to be exposed by the elements or unintentional excavation (at least with respect to finding fossils) before they can even be spotted. Other learned skills add to the successful location of fossils: a good working knowledge of geology and places likely to produce fossils, as well as the recognition of the variations in the appearance of fossils due to lighting, orientation, weathering and other damage.

Most people are familiar with the use of plaster jackets in fossil excavation: a rock-enclosed fossil is excavated and wrapped in a protective shell of plaster to safeguard it during transportation to its preparation destination. This is indeed a very common method for large vertebrate remains. But as a moment's thought will attest, vertebrates are but a ridiculous smidgen of life's history. Just like the entities preserved as fossils, the ways in which they are found and collected varies enormously.

Many fossils are found loose from their host rock, like ancient bones and shells found on a modern beach; whole brachiopods or small rocks with trilobite parts lying free on inland slopes that were once their marine mud bed; water-worn cobbles composed of the weathered colonies of corals sitting on a lake shore; or chunks of petrified wood scattered among modern desert cacti. These require no more collecting work than merely stooping to pick them up. Other fossils are dredged up off the sea floor, found by splitting concretions, sifted from stream beds or skimmed off the surface of the sea (amber pieces float in salt water). And others still are collected only virtually, by camera or sketchpad, due to their immensity, fragility, inaccessibility or ephemeral nature. Time constraints or the

11

In continental environments, animal remains tend to be scattered as these remains of a mammal in the Egyptian desert have.

legality of collecting are other reasons one might only take a picture.

Fossil preparation, whether for study or display, is sometimes thought of as an extension of collecting because digging often goes on in the lab as it does in the field, though on a smaller and more controlled scale. The most commonly illustrated technique is the mechanical removal of rock, whether with traditional hammers and chisels or with more sophisticated miniature jackhammers and carbide needles. Some types of fossils require chemicals such as acid to etch away encasing rock, or others to flocculate matrix away from microfossils. Polishing, thin sectioning and CT scanning are other lab techniques aimed at illuminating hard to reach details of certain fossil remains.

Moulding and casting are also common in palaeontology labs, replicating fossils for research or display. Exhibited specimens – whether actual fossils or high-fidelity casts – are an important way for palaeontologists to reach, teach and inspire the public.

Finally, assuring that collections of specimens are accessible to researchers is critical to the continued progress of any science. The most useful fossils are those with as much recorded data as possible, especially with respect to original geologic context. Such records ensure stability and usefulness of any information contained within the fossils for the use of future scientists. The best repositories for fossils are academic institutions, such as universities and museums, where they remain in the public domain and are thus accessible to any qualified researcher in perpetuity.

CAVEATS

As mentioned before, this book is an introduction to the many facets of palaeontology. It is hoped that this account will be as accessible to the interested layperson as to the more academically inclined. Where possible, each entry incorporates both formal and informal (like invertebrate or amphibian, which have no strict scientific usage any longer) and common names where these are available (Stegosaurus has no common name, for example); from there, the levels of information progress for each specimen.

One my note the use of 'sp' in many of the scientific names included in the pages that follow. This is a sort of place-holder abbreviation meaning 'species' and is used when the specimen shown could not be identified any more narrowly than to its genus, either due to a lack of diagnostic anatomy or because the identifier was incapable of restricting the identification any further. The use of 'isp', which signifies an 'ichnospecies' is synonymous with 'sp'. But its use is restricted to trace-fossil taxa. Lastly, some of the categories at the bottom of each page are qualified with parentheticals in order to make clear whether the data refers to the group to which the organism belongs or the individual fossil shown.

Palaeontology is a monstrously huge and varied topic. A sound attempt was made here to fly over it and point out some of that spectacular variation, and hopefully this will inspire the reader to delve deeper. The undeniable beauty of these objects – of natural art, really – is apparent, and we hope that some of their hidden magnificence has been exposed here.

Fossils are time capsules with stories that anyone can learn to read.

Ripple Marks

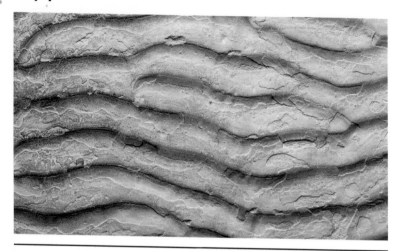

The flow of water in a river or stream, or the back and forth motion of water on a beach, leaves behind a series of ripples in the sand bed. Ripples, millions of years old, indicating water movement, have been found petrified in sandstone. Comparisons with ripple marks occurring today can reveal the direction of the current in prehistoric times, the type of environment through which the water originally flowed and even how deep the water was when the ripples formed. Although often referred to as fossils, lithified ripple marks do not qualify as such because they are not created by living things.

Class:	Not applicable
Order:	Not applicable
Habitat:	Found in many shallow aquatic environments
Distribution:	Global
Time scale:	Pre-Phanerozoic–Recent

Ozarkcollenia laminata

*O**zarkcollenia*** is a stromatolite, a structure created by the growth of colonies of photosynthesizing cyanobacteria and the attendant cementation of sediment particles during growth. Stromatolites include the most ancient fossils on Earth: some in Australia show life existed 3.45 billion years ago. Stromatolites are still forming today in seawater of extreme salinity, limited water circulation and with dissolved calcium carbonate from which the cyanobacteria cement their structure. They live in shallow clear waters with little suspended sediment in order to effectively photosynthesize. These restrictions mean they can be used to understand the paleoenvironments they are found in. *Ozarkcollenia* may have formed in a lake of volcanic origin and are dated to well over a billion and a half years.

Order:	Not applicable
Family :	Not applicable
Habitat:	Shallow marine
Distribution:	Southern USA
Time scale:	Proterozoic

Jacutophyton sp.

Jacutophyton is a type of stromatolite. The word 'stromatolite' comes from the Greek 'stroma' meaning mattress or bed, and refers to the appearance of the formations of blue-green algae that have existed for almost four billion years. Stromatolites are the only fossils we have that date back to the first seven-eighths of Earth's history, when the environment was too hostile to support any other life forms. In fact, these ancient micro-organisms played an invaluable role in shaping our planet. Stromatolites used a process called photosynthesis (a chemical 'reaction' using the light from the sun, carbon dioxide and water) to make the sugars that they fed on. A by-product of this process is oxygen, so stromatolites actually helped to make the earth habitable for oxygen-breathing life forms.

Order:	Currently under revision
Family:	Unclassified
Habitat:	Tropical regions
Distribution:	Global
Time scale:	Mesoproterozoic

Solenoporacea

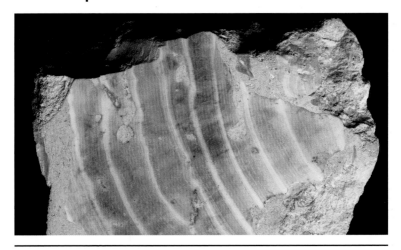

Until recently, plants were thought to be monophyletic, which means they all presumably descended from a single common ancestor and that only plants evolved from that ancestor. Recent studies suggest that 'plants' might comprise three independently evolved groups which can be called green plants, brown algae, and red algae, named so for the colour of their photosynthetic pigments. The long-lived fossil group called Solenoporacea is a group of encrusting red algae. The red pigment of the rhodophytes allows these algae to photosynthesize at lower depths than other photosynthesizers. Encrusting rhodophytes are often big contributors to reef architecture, sometimes more so than corals, and because of their similar habit of secreting a carbonate structure also go by the name coralline algae.

Order:	Cryptonemiales
Family:	Solenoporacea
Habitat:	Open marine
Distribution:	Global
Time scale:	Late Cambrian–Paleocene

Lepidodendron sp.

*L*epidodendron, often called giant club moss, grew all over the world wherever there was hot, humid swampland. The tree produced several stigmaria (thick, bristly root-like structures) in order to anchor itself firmly in shallow soil. It also produced branches, crowned at the top with leaves that resembled grass, and lepidostrobi (cigar-shaped cones). Most of the trunk remained bare, except for a covering of diamond-shaped leaf-bases that made the bark look as if it were sheathed in scales. *Lepidodendron* could grow to a height of around 40m (130ft), although 30m (100ft) was more typical. Lycopods like *Lepidodendron* had a long fossil history stretching back over 439 million years.

Class:	Lycopsida
Family:	Lepidodendraceae
Habitat:	Swamps
Distribution:	Global
Time scale:	Carboniferous

Sigillaria sp.

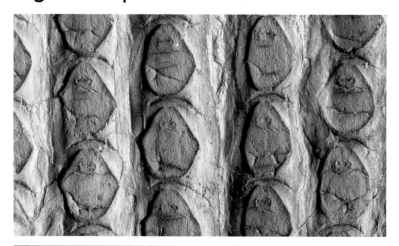

Sigillaria was a common arborescent (tree-like) lycopod. Remaining lycopods are inconspicuous forest plants that rarely top 30cm (11¾in) – in the Carboniferous, members of the group, like *Sigillaria*, reached heights of 40m (130ft). The arborescent lycopods were unlike modern tree-sized plants. They had photosynthetic trunks, evidenced on this Carboniferous specimen by the numerous organized leaf scars. Most branched only at maturity in order to produce cones for reproduction: at earlier life stages, they were unbranching trunks covered in scale-like leaves. Also, they had heavy, symmetrically arranged 'roots' called stigmarian rhizomorphs, or stigmaria with numerous radiating rootlets, which some scientists believe were photosynthetic, too, and helped the photosynthetic rhizomorphs reach sunlight in flooded swamps.

Class:	Lycopsida
Family:	Sigillariaceae
Habitat:	Swamps
Distribution:	Global
Time scale:	Carboniferous–Permian

Annularia sp.

*A**nnularia** is the name given to fossils of certain ancient horsetail plants. They can be told apart from other horsetails by the patterns made by their leaves. *Annularia* had soft leaves that formed star-like rings around the stem. The base of each leaf in one of these rings came from the same area of the stem. Other horsetails had rigid leaves, which arched upwards rather than sticking straight out from the sides. *Annularia* grew in damp places such as lake shores, swamps and the banks of rivers. It existed during the Carboniferous Period. This was the time when most of the world's coal was formed and buried. (Coal is the fossilized remains of ancient plants.)

Class:	Calamitales
Order:	Calamitaceae
Habitat:	Swamps & the banks of lakes & rivers
Distribution:	Warm regions globally
Time scale:	Carboniferous–Permian

Annularia sp.

The leaves of *Annularia*, a sphenopsid, developed from circles surrounding the stem of *Calamites* plants. *Annularia* was generally arborescent (tree-like) and could grow to a fair height – around 10m (30ft). As fossils, they have often been found impressed onto the slabs of coal that developed from the trees of the Carboniferous Era. *Annularia* enjoyed worldwide distribution but was particularly abundant in Europe.

Class:	Calamitales
Order:	Calamitaceae
Habitat:	Swamps & the banks of lakes & rivers
Distribution:	Warm regions globally
Time scale:	Carboniferous–Permian

Calamites sp.

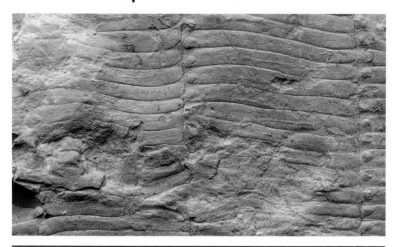

Although they are related to today's horsetails, *Calamites* actually looked more like a tree. Growing to around 9m (30ft) tall, their branches typically grew in concentric circles that spiralled up and around the trunk. Sword-shaped leaves and cones were arranged in whorls along each branch. Like bamboo, the trunks of *Calamites* were hollow. When the plant died, the hollow section of the tree often filled with mud and sediment. Over time, the exterior decayed, leaving a cast impression of the plants central core behind. These casts are perhaps the most common form of *Calamites* fossil, although other well-preserved tissues are also found.

Order:	Calamitales
Family :	Calamitaceae
Habitat:	Warm, wet regions
Distribution:	Global
Time scale:	Carboniferous–Permian

Sphenopteris sp.

Fossil plants are almost never found complete because of the fragility of many plants and the giant size of some. Also, many plants evolved to detach parts of their anatomy at certain points in their lives, like cones, seeds or leaves. As fossils, it is common to find only petrified wood, or only seeds, and other times only leaves. The fact that sediments generally only preserve certain types of plant parts contributes to this situation. *Sphenopteris* is a genus of fossil leaf usually considered to be from a true fern. True ferns are spore-bearing plants whereas 'seed ferns' are an extinct group of plants with superficially fern-like leaves that reproduce by seed. Unless *Sphenopteris* leaves associated with reproductive structures are found, it is hard to be sure if it is a true fern or not.

Class:	Filicopsida
Family:	Not applicable
Habitat:	Swamps & forests
Distribution:	Global
Time scale:	Devonian–Cretaceous

Tietea singularis

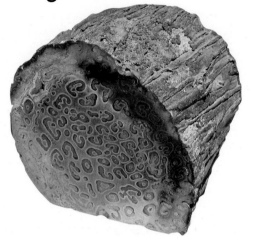

Tree ferns are an ancient lineage of plants that extend all the way back to the Carboniferous Period. Botanically, they are not considered trees since their stems are not composed of true wood. Instead, their trunks are formed by leaf bases extending upwards and fine roots winding downwards. They could reach a considerable height and are still common plants of the tropics and other warm land habitats. The Permian *Tietea singularis* is very common in the Pedro de Fogo Formation of Brazil, where it is exquisitely preserved. The silica minerals that have replaced the plant's tissues faithfully reproduce the structure down to the cells. Anatomically preserved plants like this – as opposed to plant fossils that have been, for instance, carbonized – are often the most palaeobotanically useful.

Class:	Marattiopsida
Order:	Marattiales
Habitat:	Wet forests
Distribution:	Brazil
Time scale:	Permian

Pecopteris unita

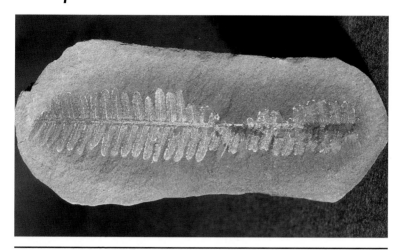

*P*ecopteris is the name given to fossil fern leaves that grew from the top of trunks found as fossils that hold the name *Psaronius*. For a proper identification, *Psaronius* stems need to have been petrified in order to preserve the diagnostic anatomy. The leaves, however, are commonly preserved in rocks that were formed in more general conditions, thus often being the only evidence that *Psaronius* trees were around. Psaronius was a tree fern that reached heights of 15m (50ft) and, like all tree ferns, had a mantle of tiny roots that reached down the outside of the trunk to the ground called adventitious roots. This root mantle was often home to epiphytic ferns, which are ferns that grow on other plants, as well as vine-like ferns and seed ferns.

Order:	Marattiales
Family:	Marattiaceae
Habitat:	Wet lowlands, peat forming wet woodland & dry grounds
Distribution:	Global
Time scale:	Carboniferous

Osmundacaulis sp.

Most ferns today are ground plants familiar in moist woods. And although tree ferns are still around today, they were much more common in the distant past. This is a polished cross-section of the stem of *Osmundacaulis*, an osmundaceous fern from the Wandoan Valley of Queensland, Australia. It shows the detail separating these plants from true trees, which have a woody stem: the central stele with radiating bundles of vascular tissue, and the double-spiraling leaf bases that make up the bulk of the stem. *Osmundacaulis'* family is often referred to as the 'flowering ferns' because of the unusual appearance of their terminal, ripe sporangia or reproductive structures. Today's osmundaceans tend to have short trunks but those found in the fossil record show certain species reached tree-like proportions.

Order:	Filicales
Family:	Osmundaceae
Habitat:	Wet forests
Distribution:	Australia
Time scale:	Jurassic

Alethopteris scalariformis

In the early years of palaeobotany, *Alethopteris* and its allies were regarded as ferns because of their foliage's striking similarity to the leaves of many pteridophytes, the group to which ferns belong. But as better fossils were unearthed, it became clear that these superficially fern-like plants possessed a very different mode of reproduction. Rare specimens were found showing that they grew from seeds, as opposed to spores like ferns do. A plant's reproduction is very important in determining its classification and *Alethopteris* and others belonged in a new group, named pteridosperms, informally dubbed the 'seed ferns'. The idea that they were something unique was tentative, but as even more fossils confirmed this case, the seed ferns became a well-established group.

Order:	Medullosales
Family:	Medullosaceae
Habitat:	Swamps
Distribution:	Global
Time scale:	Carboniferous

Cordaites sp.

The late Paleozoic tree *Cordaites* was part of a now-extinct group of gymnosperms with strap-like leaves like the carbonized foliage seen here. It took decades to come to an understanding of what the tree looked like and how it reproduced and so different parts of *Cordaites*' anatomy have different names. Infillings of the central cavity are known as Artisia. The cone-like reproductive parts are called *Cordaianthus*. Branches are found with visible leaf bases are *Cordaicladus*. The pollen is of the grain type *Florinites*. Heart-shaped, flattened seeds attributed to *Cordaites* are *Cordaicarpus*, *Cardiocarpus*, and *Samaropsis*. Its wood is called *Cordaixylon* if wood, bark and pith are present. Fossil wood called *Dadoxylon* also originates from *Cordaites*. And root wood is called Amyelon.

Order:	Cordaitanthales
Family :	Pinopsida
Habitat:	Various including mangrove-like coastal to dry terrestrial
Distribution:	Global
Time scale:	Carboniferous–Permian

Ginkgo (genus) sp.

Commonly known as the 'maidenhair tree', ginkgos are a rare example of a 'living fossil'. This once wide and diverse group of plants appeared on Earth 270 million years ago and one species, *Ginkgo bilboa*, survives to this day. These tall deciduous trees are surprisingly hardy. Indeed, *Ginkgo* trees in Hiroshima survived the atomic blast when almost everything else in the immediate area died. Easily recognized by its distinctive fan-shaped foliage, this Palaeocene ginkgo leaf comes from the Sentinel Butte Formation of North Dakota, USA. The seeds of *ginkgos* are unusual too. Most trees protect their seeds inside fruit, but Ginkgos are gymnosperms. The name means 'naked seed', so while these ancient trees do produce round 'berries', these are not actually fruit.

Order:	Ginkgoales
Family:	Ginkgoaceae
Habitat:	Tolerant of wide range of habitats
Distribution:	Global
Time scale:	Permian–Recent

Neuropteris sp.

*N*europteris, which flourished chiefly in Europe, North America and Asia, comprised the foliage of a medullosan, an extinct seed-bearing plant. Like *Pectoperis*, *Neuropteris* has been classed as an 'artificial' rather than a 'natural' genus because its species have not been properly defined as a group. *Neuropteris*, for example, was put together with other species on one basis – its venation (the pattern of the veins on its leaves) was similar to that on the leaves of other plants placed in the same genus. *Neuropteris* had a single vein along the centre of its large leaflets, and this vein either forked into a fan of smaller veins or carried on as a single vein which, on reaching the edge of the leaflet, gave it a fan-shaped curve.

Class:	Gymnospermae
Order:	Pteridospermae
Habitat:	Wet Forests
Distribution:	Global
Time scale:	Early Carboniferous–Early Permian

Glossopteris sp.

Glossopteris is popularly known as the Gondwana tree after Gondwanaland, the southern part of the single continent Pangaea which split off from the northern part, Laurasia, about 200 million years ago. Consequently, its fossils have been found mainly in the southern hemisphere. *Glossopteris* could reach a typical height of 8m (26ft), and grew rosette-shaped leaves that could become very large. One *Glossopteris* fossil, found in India, had broad sword-shaped leaves. The tree grew best in warm, damp lowlands. It bore specialized leaves that revealed fruit-bearing propensities, such as seed-bearing structures and pollen capsules. *Glossopteris* wood was softwood. The rings in evidence on the fossil *Glossopteris* probably grew in response to seasonal changes in the climate.

Class:	Pteridospermopsida
Order:	Glossopteridales
Habitat:	Various terrestrial
Distribution:	Southern continents (Gondwana)
Time scale:	Permian

Cycadeoidea sp.

Cycadophyta is a once-large group of plants that exists today as the remaining subgroup called cycads. The group's heyday was the Mesozoic when the other subgroup, called cycadeoids, enjoyed more diversity and extended range. The two groups were superficially similar with a stem composed externally of spirally arranged leaf bases and a crown of rigid finely-pinnate leaves. But the main difference was in the arrangement of the reproductive structures. True cycads wore their cones at the top of their often squat stems, whereas cycadeoids dispersed their cones throughout the tough armor of their scaly leaf bases. The genus showed here as a partial specimen of the stems' exterior is *Cycadeoidea*, which had sparse cones, but the genus *Monanthesia* had a cone for every leaf base.

Order:	Bennettitales
Family:	Not applicable
Habitat:	Forests
Distribution:	Global
Time scale:	Early Permian–Cretaceous

Walchia filiciformis

The Norfolk Island pine in Australasia and the Monkey Puzzle Tree are present-day examples of Araucaria, an ancient group of large tree originating during the Triassic Period. *Walchia* is a genus of fossil plant that at least in part probably represents an araucarian. A difficulty with fossil plants is that they rarely become preserved whole, or at least with diagnostic reproductive organs associated. *Walchia Culmitz*, *Hermitia* and *Ernestiodendron* are all fossil plants from genera (see *Palmoxylon*) named for the presence or absence of certain parts. They may all actually be the same plant preserved differently.

Class:	Pinopsida
Order:	Araucariaceae
Habitat:	Terrestrial
Distribution:	Global
Time scale:	Late Carboniferous–Permian

33

Araucaria mirabilis (cone)

Araucarian trees, or trees closely related to them, have an extensive history reaching back to the Triassic. These trees dominate the Triassic palaeoflora of Petrified Forest National Park in Arizona, USA. But it is in Cerro Cuadrado, Patagonia that the most spectacular fossils of this group are found. Volcanic sediments that killed and buried Jurassic araucarians not only preserved their trunks and twigs in exquisite detail, but also their cones. Cerro Cuadrado's are the best known fossil conifer cones and for good reason: they faithfully preserve all the internal structures, down to seeds and cells, when cut and polished. Seen here is a complete, uncut cone. The scales are readily visible on the cone's exterior.

Order:	Pinopsida
Family:	Araucariaceae
Habitat:	Terrestrial
Distribution:	Global
Time scale:	Jurassic–Recent

Pararaucaria patagonica (cone)

Fossil cones or other reproductive parts of plants are useful in understanding the evolution of plants. One area famous for fossil cones is the Jurassic sediments of Cerro Cuadrado near Santa Cruz, Argentina. Cones by the thousands have been collected there, silicified by volcanic sediments in such detail that the seeds and cone anatomy within are clearly visible under magnification. Araucarian trees, are the main fossil cone producers in the area. The most common araucarian cones in Cerro Cuadrado are tangerine-sized and spherical (see *Araucaria mirabilis*), but the cones of *Pararaucaria* are much more rare, exhibit a short cigar-like morphology, and might represent a conifer of non-araucarian type.

Order:	Pinopsida (not confirmed)
Family:	Araucariaceae (not confirmed)
Habitat:	Terrestrial
Distribution:	Argentina
Time scale:	Jurassic

Arizona Petrified Forest

Millions of years ago, the trees in the Petrified Forest National Park, Arizona, became waterlogged after dying and falling. Gradually, minerals in the water silicified the original wood tissue, replacing it with silica. The result is a vista of prehistoric trees turned to stone (the word 'petrified' comes from the Latin 'petra', meaning 'rock'). Fresh finds are made every year in the Forest. Already more than 150 different species of fossil plants have been discovered, together with reptiles such as the armadillo-like *Desmatosuchus*. The adjacent Painted Desert, so-called because of the minerals and decayed plants and animals that give it brilliant colours, also has interesting fossils, notably an entire conifer forest dating from the Triassic Era.

Class:	Pinopsida (mostly)
Order:	Araucariaceae (mostly)
Habitat:	Terrestrial
Distribution:	Arizona
Time scale:	Triassic

Cupressus sp. (cone)

Cypress trees have been around at least as far back as the Jurassic and are still with us today. These rugged trees can be seen living along the west coast of North America, where high winds off the Pacific contort their branches. Because of their adaptation to windy areas, they are often planted ornamentally as windbreaks. The cones of cypress trees are common fossils of the Late Cretaceous Hell Creek Formation in western North America. The Hell Creek is famous for its dinosaur fauna, which includes *Tyrannosaurus*, but also had a large variety of herbivores like *Triceratops* and several species of hadrosaurs (duck bills). The twigs of the cypress, crowded with tiny scale-like leaves, were a likely food source for many of these plant eaters.

Order:	Pinopsida
Family:	Cupressaceae
Habitat:	Terrestrial
Distribution:	Global
Time scale:	Jurassic–Recent

Metasequoia sp.

Also known as the dawn redwood, *Metasequoia* is one of those organisms first known from fossils but then found living today, although it was only a few years that separated the naming of the fossils and the discovery of the living trees. *Metasequoia* was first described from fossils in 1941. In 1944, a small stand of unusual trees was discovered in China but, due to World War II, were not studied until 1946. In 1948, the trees were described as the only living species of dawn redwood. This Oligocene specimen comes from the Muddy Creek Formation of Beaverhead County, Montana and probably represents a shed leaf. Deciduousness is one of the features that separates this genus from the related genus *Sequoia*, which has bark and foliage very similar to the dawn redwood.

Order:	Pinopsida
Family:	Cupressaceae
Habitat:	Temperate forests
Distribution:	Global
Time scale:	Cretaceous–Recent

Angiosperm (leaves)

Angiosperms (flowering plants) are the most common and familiar plants of today. Their origins are obscure but have been around at least since the Early Cretaceous and quickly diversified into a wealth of habitats. Two main groups exist within the angiosperms: the monocots, which have parallel leaf veins and include plants like grasses and palms, and the dicots, which have branching leaf veins and include poison ivy, fruit trees and oaks. The fossil leaves pictured here show the complex venation of the dicots and are preserved as carbon films. Fossil leaves that are merely impressions are the same colour as the matrix. Carbonized leaves are formed when all the volatile compounds are driven off, leaving a trace of original carbon in the form of the original.

Order:	n/a
Family:	n/a
Habitat:	Varied terrestrial & aquatic
Distribution:	Global
Time scale:	Early Cretaceous–Recent

Palmoxylon sp.

Palms are common plants of the tropics and diversified early in the history of flowering plants. *Palmoxylon* is what is known as a 'form genus'. Since plants are very rarely found as complete fossils, they are often given different names for each part, leaving some plants with separate form genera for their wood, leaves, cones, etc. They will often keep their multiple names even when specimens are found uniting more than one form genera into one plant. One reason for this is that, occasionally, a form genus is found to belong to more than one species of plant. '*Palmoxylon*'means 'palm wood' and is used to designate stem fossils that resemble modern palm stems. Palms do not have true wood but instead their trunks are made up of bundles of fibres.

Order:	Arecales
Family:	Arecaceae
Habitat:	Coastal Terrestrial
Distribution:	Global
Time scale:	Cretaceous–Recent

Typha sp.

The genus *Typha* is the water-loving plant more commonly known as the cattail. Its tall thin form with its sword-like leaves and distinctive sausage-, or 'cat tail-', shaped flower are a common sight on pond edges and other moist areas. Most fossils of *Typha* are compression fossils that only show the flattened outlines of the leaves. But some, like these Miocene examples from the Mojave Desert in California, are true petrifactions. Seen in cross-section, they display the fully three-dimensional, concentrically layered leaves that form the main body of the plant, much like celery. Petrifaction of relatively soft plants like these is much rarer than for harder woods because the trees can survive the abuse of transport and burial better. These *Typha* were likely buried in their life position.

Order:	Poales
Family:	Typhaceae
Habitat:	Almost anywhere soil remains wet, saturated, or flooded
Distribution:	Global
Time scale:	Late Cretaceous–Recent

Juglans tephrodes (nut)

This Late Pliocene or Early Pleistocene butternut comes from a quarry along the Maas River in Arcen, Germany. It has not become mineralized but has instead carbonized. Carbonization is when all of the volatile chemicals in organismal remains are driven off, leaving a ghostly carbon remnant. The flesh of ichthyosaurs, flimsy plant parts, and even delicate feathers have been preserved this way. Carbonized remains are very fragile and are especially susceptible to damage due to changes in moisture levels, which causes them to crumble to dust. This fossil nut was brought up from 26m (85ft) below ground level by quarry dredges. This fossil is actually the same species as the extant butternut *Juglans cinerea*, but the rules of taxonomy require that the species be given a different name.

Order:	Juglandales
Family:	Juglandaceae
Habitat:	Temperate forests
Distribution:	Northern Hemisphere
Time scale:	Cretaceous–Recent

Celtis occidentalis (seeds)

These are seeds of the hackberry and can be found in great numbers in the Oligocene Orella Formation, in the White River Badlands of South Dakota, North America. Today, the local environment where the fossil seeds are found is arid and vegetation–poor. But we can tell from things like hackberry seeds that the Oligocene was more temperate, like the modern environments that the hackberry favours. Hackberry leaves are also found from this time period, but the durable seeds had an even better chance of being preserved than the comparatively fragile leaves. In some fossil mammal localities, countless seeds are retrieved whereas the leaves are scarce. They are sometimes so common that in certain areas of Nebraska the fossil seeds make up a prominent portion of the rocks.

Order:	Urticales
Family:	Ulmaceae
Habitat:	Humid to arid environments in tropical and temperate regions
Distribution:	Northern Hemisphere
Time scale:	Paleocene–Recent

Petrified Forest of Lesvos (Lesbos)

The petrified forest on the Island of Lesbos was first reported in 1844. The volcanic sediments of the area include abundant ash – the best sediment in which to petrify wood. This Protected Natural Monument exhibits preserved leaves, cones and seeds, but it is the trees that are most impressive. Many of them are still standing upright as when they were alive millions of years ago. Fossils like this are called autochthonous, which means they remained exactly where they lived. Most other petrified wood comes from trees that were transported elsewhere after falling, for example, by a swollen river. But the trees of Lesbos were killed and buried in place by their preserving volcanic sediments. Many species of plants have been identified here and show a subtropical environment with mild temperatures.

Order:	Several
Family:	Several
Habitat:	Subtropical forest
Distribution:	Greece
Time scale:	Late Oligocene–Middle Miocene

Petrified Forest of Egypt

Strewn over a vast area of the Libyan Desert, especially the desert basin known as the Qattara Depression of north-western Egypt, is an enormous amount of petrified wood. It is not uncommon to find whole trees, like the one shown here, incongruously laid out on the desert floor. This mismatch clearly points to a wetter past for this now incredibly arid region. One early visitor to the area in the eighteenth century, after realizing that the translation of the area's name meant 'sea without water', assumed that these trees were the petrified masts of ships and thus proved a historic time when a sea covered this land. The area was indeed wetter in the past, but this wood can claim a much older heritage: the trees preserved here are around 30 million years old.

Order:	Several
Family :	Several
Habitat:	Tropical forest
Distribution:	Egypt
Time scale:	Oligocene Period (not confirmed)

Petrified wood

Woody trees have a long history going back to at least the Devonian and have lived in nearly all terrestrial environments. The conditions that are necessary to petrify wood are also very common, making petrified wood a common fossil in many places on Earth. This small chunk of petrified wood comes from Banks Island in arctic Canada. The current tundra environment in which this piece was found does not allow tree growth because of permafrost just underground prohibiting root penetration necessary for tall plants. But this specimen proves a different climate for Banks Island's past. The problem is that fossils like this, found loose on the surface, are out of their geological context, making reliable dating impossible. This wood could have come from anywhere in the more than 400 million years of wood.

Order:	n/a
Family:	n/a
Habitat:	Forests
Distribution:	Global (wood)
Time scale:	Carboniferous–Recent (wood)

Polyporites wardii

We know that fungi go very far back in the fossil record despite their very poor representation as fossils. There are a couple of mushrooms in amber and some possible petrified specimens, but by far the most commonly preserved fossil fungi occur in petrified wood and appear as tiny fungal filaments called hyphae. The damage caused by fungi can also be recognized in many fossil woods, such as this piece of *Araucarioxylon* from Triassic Chinle Formation of southern Utah. All along the edge are pockets of rot caused by the fungus *Polyporites wardii*. Not all fungi are parasitic, and many have symbiotic relationships with trees and other plants crucial to both plant and fungi. These relationships are widespread among modern plants and probably evolved early in the history of plants.

Order:	Agaricales
Family:	Polyporaceae
Habitat:	Forests
Distribution:	Southwest USA
Time scale:	Triassic

Actinostroma verrucosa

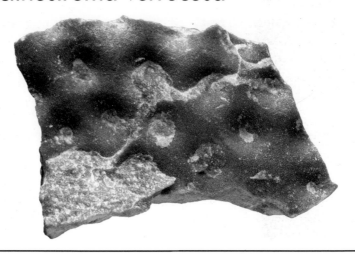

*A*ctinostroma is a stromatoporoid. There is little in their structure that links this group securely with known groups. They were clearly reef-building colonial forms with skeletons of calcium carbonate, but so are many other unrelated groups, like some algae and many corals, with which they do share some features. This specimen shows the internal contours peculiar to the group, which are exposed upon weathering. Many scientists now place the stromatoporoids within the Porifera, or sponges. Since many groups of organisms have become extinct long ago, leaving us nothing but fossils which often lack crucial diagnostic features, it is no wonder that so many remain mysterious. But like so many other groups once baffling, stromatoporoids may one day reveal themselves as better fossils are found.

Class:	Demospongia
Order:	Stromatoporoidea
Habitat:	Marine
Distribution:	Global
Time scale:	Late Ordovician–Late Devonian

Spriggina floundersi

In common with *Dickinsonia costata*, *Spriggina floundersi* is a 'mystery fossil' from the Pre-Cambrian. In the past, it was suggested that *Spriggina* were plants or lichens rather than animals. However, most palaeontologists now agree that *Spriggina* fossils show the remains of a creature with a segmented body and a clearly identifiable head and tail. Some examples even seem to show a circular mouth at the centre of a horseshoe-shaped head. *Spriggina* was once classified as a type of marine worm, although it has now been suggested that it may be an early ancestor of the trilobite. *Spriggina floundersi* takes its name from R.C. Sprigg and Ben Flounder, the geologists who discovered it in 1951.

Order:	Disputed
Family:	Disputed
Habitat:	Marine
Distribution:	Australia
Time scale:	Ediacaran

Dickinsonia costata

Our Earth has changed so much since life began that it can sometimes be hard to interpret what we find in the fossil record, and the further back in time we go, the harder it becomes. During Ediacaran times, 600 million years ago, Earth was a truly alien planet. This makes categorizing fossils based on a comparison with today's known organisms almost impossible. *Dickinsonia costata* is one of these 'enigma fossils'. Although it resembles an annelid worm (such as a leech), it may equally be some unknown form of jellyfish, coral, lichen or a species completely new to science. Fine examples of *Dickinsonia* fossils have been found throughout Australia, a fact celebrated in 2005 when it featured on an Australian Post 50 cent stamp.

Order:	Dickinsoniida (disputed)
Family:	Dickinsoniidae (disputed)
Habitat:	Marine
Distribution:	Australia and White Sea region of Russia
Time scale:	Ediacaran

Charniodiscus oppositus

These soft-bodied invertebrates lived about 600 million years ago. That is about 100 million years before the arrival of the first fish, 370 million years before the first mammals and about 590 million years before the first recognizable human. *Charniodiscus* was a 'filter feeder', attaching itself to the sea bed with a bulbous shaped 'holdfast'. This allowed it to gather up organic material without being swept away by the current. At least five species of *Charniodiscus* have so far been identified and each one can be distinguished by the number of segments into which their leaf-shaped bodies are divided. Little detailed information is known about *Charniodiscus oppositus*, as few complete fossils have ever been found.

Order:	Rangeomorpha
Family:	Charniidae
Habitat:	Marine
Distribution:	Australia, Canada, Russia and UK
Time scale:	Ediacaran

Invertebrate (bioturbation)

Trace fossils are some of the earliest evidence of animals in the fossil record. Since the first animals lacked hard parts, we often have only the marks they left as evidence of their presence. These are from the sandy seabed of about 500 million years ago in a sediment that is unlikely to have preserved the bodies of soft animals. Accumulations of traces like this, where the animals left marks on nearly the entire surface they traversed, are referred to as bioturbation. The movements of individual animals tend to be obscured in fossils such as these, but an impression of abundant life is apparent. Comparison with modern animal behaviour in a similar environment as the one in which these traces were made can help interpret the ancient trace-makers' behaviour.

Class:	n/a
Order:	n/a
Habitat:	Marine sand
Distribution:	St. Lawrence River, New York, USA (this specimen)
Time scale:	Late Cambrian–Early Ordovician (this specimen)

Siphonia sp.

Siphonia belonged to a genus of European demosponges. Sponges were many-celled organisms with a body that was built around the needle-like spicules of its skeleton. It was oblong in shape and became wider as it approached the base. The surface looked smooth, but on closer acquaintance revealed a mass of tiny pores. The long and narrow stalk which supported the main body of the *Siphonia* sponge gave it the appearance of a mushroom or a tulip. The sponges are very primitive animals and lack true tissues and thus have no muscles, nerves or internal organs. They merely circulate water throughout their bodies to extract the necessary nutrients.

Class:	Desmospongia
Order:	Tetralithistida
Habitat:	Marine
Distribution:	Europe
Time scale:	Middle Cretaceous–Early Tertiary

Raphidonema farringdonense

Sponges such as *Raphidonema farringdonense* are one of the most primitive forms of aquatic life. Most are 'sessile', meaning that they attach themselves to a rock and move very little (or not at all) during their lives. Sponges have no muscles and no nervous system. Their bodies are simple sacs that draw water and food in through small holes and expel waste through larger holes. Tough, silica-based 'skeleton' formations support the body. *Raphidonema farringdonense* belongs to a separate class of sponges that have calcareous skeletons and a rough cup-shaped exterior. Sponge fossils are hard to find, although they occur in abundance in the Farringdon Sponge Gravels, in the UK, which is where *Raphidonema farringdonense* was discovered.

Order:	Pharentrondia
Family:	Lelapiidae
Habitat:	Warm, shallow waters
Distribution:	Most famously found in the Faringdon Sponge Gravels, Oxfordshire, UK
Time scale:	Early Cretaceous

Discodermia sp.

Discodermia was a type of sponge that lived in the oceans around 100 million years ago, when non-avian dinosaurs still populated the earth. Sponges are among the most simple and primitive of all animals. The earliest sponges appeared well over 500 million years ago, long before there were plants or other complex life forms on land. *Discodermia* was a filter feeder, like all other sponges before and since. It lived on tiny animals and other particles of food sieved from the water which it sucked in through the holes in its surface, then pumped out once everything edible had been removed. Close relatives of this particular sponge still exist today in the deep sea.

Order:	Choristida
Family:	Discodermiidae
Habitat:	Ocean floor
Distribution:	Global
Time scale:	Cretaceous

Nummulites sp.

Foraminiferans were protozoa – basic, primitive organisms whose nature was reflected by their original description: the word 'protozoa' comes from the Ancient Greek for 'first' (proto) 'animal' (zoion). They were one-celled organisms with skeletons composed of calcium carbonate, silica, organic material, or of grains of sand fused together like a suit of armour. Foraminiferans could be microscopic, measuring only 0.05mm (0.001in) to 10cm (4in). *Nummulites* lived in the Mediterranean Sea and east Caribbean Sea areas and were one of the larger members of the Order. Their skeletons were composed of calcum carbonate perforated with small holes. *Nummulites*, extinct since the Oligocene Period, were decorated with curves or spirals, sometimes accompanied by tiny nodules.

Class:	Granuloreticulosea
Order:	Foramniferida
Habitat:	Marine
Distribution:	Western Hemisphere
Time scale:	Palaeocene–Oligocene

Astrangia lineata

This stony coral closely resembled several species that still live in the North Atlantic Ocean today. Like them, it formed small incrustations on rocks and other hard objects rather than the massive reefs built by more tropical species. *Astrangia lineata* fed on planktonic animals and other tiny particles of food in the water, capturing them with miniature stinging tentacles. It also contained single-celled algae in its body tissues, which generated extra food in the form of sugars created by photosynthesis (a process that harnesses the energy of sunlight). *Astrangia lineata* laid down a protective limestone skeleton as it grew, and it is this that is preserved as the fossil. Several of the soft-bodied polyp animals lived together in colonies, each one occupying its own circular flower-like structures.

Order:	Scleractinia
Family:	Astrangiidae
Habitat:	Shallow seas
Distribution:	North America
Time scale:	Miocene

Pleurodictyum *problematicum* and *Hicetes* sp.

***P**leurodictyum* was not a single creature but a colony formed by polyps closely pressed together. The colony was not particularly large – only about 1.7cm (¾in) in diameter – but in fossil form, it has been found over a wide-ranging area, in Europe, Asia and Africa. The walls of the colony were thick and strong and contained a large number of pores. The actual septa (interior dividing walls) were fairly primitive and were shaped like spines. The tubes of the worm *Hicetes* were often found together with those of *Pleurodictyum,* and the two seem to have had a symbiotic relationship. *Hicetes* was a tube-forming worm and lived within the *Pleurodictyum* colony, fusing its tube with the coral's exoskeleton.

Class:	Anthozoa
Order:	Favositida
Habitat:	Marine
Distribution:	Europe, North America
Time scale:	Devonian

Caryophyllia sp.

Caryophyllia is a cup-coral. Corals in cold waters do not build reefs like their tropical kin do. Instead, they make individual limestone cups in which to live, as opposed to the fused colonial structures of tropical reef corals. The free-living larvae find rocks on which to cement their cups, after which they spend the rest of their lives often in small groups. The modern species of *Caryophyllia* are widely dispersed in European coastal marine waters and can usually be found on vertical or overhanging rock faces and sometimes in small caves out of the light. The Pliocene specimens here show the septa dividing the cup's mouth and the attachment point at the opposite end. The specimen on the right shows the curved tunnel of another organism that bored through the coral after it died.

Order:	Scleractinia
Family:	Caryophylliidae
Habitat:	Rocky coastal marine
Distribution:	Global
Time scale:	Early Jurassic–Recent

Montlivaltia sp.

*M*ontlivaltia are unusual corals in that they are solitary and unattached to a colony of corallites. *Montlivaltia* are quite large and often cylindrical in form. There are numerous septa that have indented edges. These septa touch each other and by this means form a long groove. Often *Montlivaltia* displays no epitheca (the continuous layer covering most of the exterior of many corals). *Montlivaltia's* columela is more often than not only primitively developed, with few spiral thickenings. *Montlivaltia* has been found in fossil form from the Middle Triassic to the Pliocene. *Montlivaltia* corals can be tall: *Montlilvaltia lessneuri*, for example, measured 7cm (2¾in) high.

Class:	Anthozoa
Order:	Scleractinia
Habitat:	Marine
Distribution:	Global
Time scale:	Triassic–Pliocene

Hexagonaria sp.

*H*exagonaria was a colonial coral comprising huge colonies in the shape of loaves. Between the corallites (the individual skeletons of the colony's inhabitants) the walls were very strong, and these were in contact all along the length of the side walls of *Hexagonaria*. As the colony's name suggests, the cross section tended to be six-sided. *Hexagonaria* had many septa (thin dividing walls) and these were usually long. In the Devonian, *Hexagonaria* was common the world over, and today its fossils are often found along the shores of Lake Michigan and Lake Huron in the United States. Some 350 million years ago, during the Devonian, *Hexagonaria* lived in the shallow warm seas that covered the state of Michigan.

Class:	Anthozoa
Order:	Rugosa
Habitat:	Marine
Distribution:	Global
Time scale:	Ordovician–Carboniferous

Halysites sp.

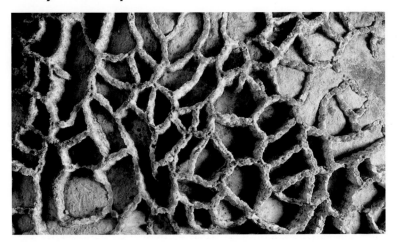

*H*alysites was a colonial coral with a 'pan-pipe' formation of corallites (individual skeletons) in a single series. There were tiny tubes between the round or elliptical corallites forming a chain-like cross section. These chain forms vary – some are straight, some are curved. These divided and rejoined, so forming cell-like spaces in the process. There are no septa (thin dividing walls), although some of these appear as a scattering of spines. *Halysites'* walls, by contrast, were thick and there were numerous tabulae (the transverse walls that enclosed the bottom of the cavity in which individual colonial creatures lived). *Halysites* lived worldwide, in warm shallow waters and reefs.

Class:	Anthozoa
Order:	Halysitida
Habitat:	Marine
Distribution:	Global
Time scale:	Ordovician–Silurian

Calceola sp.

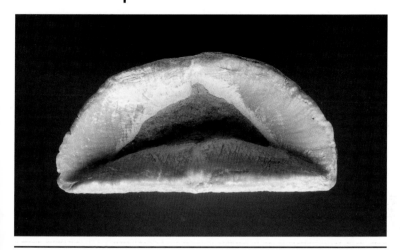

As the oldest genus of Paleozoic corals still in use, *Calceola*, not surprisingly, has a tangled nomenclatural history, having changed names several times. And since much of this was done in the early years of naming in biology, before most of the kinks were worked out, current workers are left with a confusing and cumbersome history to learn before moving forward. In addition to these problems, *Calceola* went through a number of different interpretations as to its identification. At times, it was seen as either a rudist bivalve (like *Diceras* and *Hippurites*) or a brachiopod. This is not unreasonable considering that both of these groups have coral-like form with lid-like valves covering a cone shaped one.

Now, it is very clear that it was a bizarre, lidded coral.

Order:	Anthozoa
Family:	Rugosa
Habitat:	Marine
Distribution:	Europe, Africa, Asia, Australia
Time scale:	Devonian

Essexella sp.

Fossils of soft-bodied creatures are rare, and there are few animals with softer bodies than jellyfish. *Essexella* was a jellyfish that lived in the oceans which covered parts of what is now North America around 300 million years ago. Its fossils are usually oblong, showing the flattened 'bell' it used for swimming. In life, this would have trailed a long 'cape' of tentacles covered with stinging cells, which it used to capture its prey. *Essexella*, like all jellyfish, was a primitive animal with a relatively simple body structure. While it was drifting in the seas, the land of North America was covered by forests and swamps inhabited by insects and the earliest amphibians.

Class:	Scyphozoa
Order:	Rhizostomatida
Habitat:	Open ocean
Distribution:	North America
Time scale:	Carboniferous

Conularia crustula

Conularids are bizarre animals that lived through most of the Palaeozoic and Mesozoic. *Conularia crustula* was a Carboniferous form and a fairly typical member of the group. For over 150 years, scientists have been trying to work out the classification of these animals with little luck. Many presume them to be some kind of scyphozoan (jellyfish) because of their fourfold symmetry, but other of their features are utterly unlike any known scyphozoan; elongated, pyramidal exoskeletons made up of rows of calcium phosphate rods. In rare examples where potentially helpful soft parts have been preserved, they still offer little to clarify the group's position. Some reconstructions have them floating with their points up swimming like jellyfish, and others have them attached to hard substrates by a flexible stalk.

Order:	Conulariida
Family:	Conulariidae
Habitat:	Marine
Distribution:	Global
Time scale:	Carboniferous

Ottoia prolifica

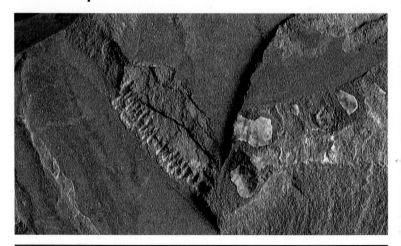

Fossils of these cylindrical worm-like creatures are found in abundance in Canada's Burgess Shale deposits. Growing up to 8cm (3¼in) in length, *Ottoia prolifica* is believed to have been an active burrower, spending much of its time hidden beneath the sediment, waiting for prey to pass by. Using a long proboscis to catch its victims, *Ottoia* seems to have been a voracious hunter. Fossils have even been found with other undigested priapulid worms in their guts, suggesting that this species occasionally resorted to cannibalism. Although *Ottoia prolifica* was probably not very mobile, its body was filled with a fluid which acted like a skeleton, allowing it to locate.

Phylum:	Priapulida
Family:	n/a
Habitat:	Seabed
Distribution:	Fine examples found in Canada's Burgess Shale
Time scale:	Cambrian

Worm (casting)

Many marine worms tunnel through muddy sediments, ingesting them and extracting the organic matter contained within for nutriment. After feeding like this for a time, the indigestible particles of sand are ejected outside of the burrow at the surface and remain as a strand of worm-shaped sediment near or on the burrow exit. This Miocene fossil example from Switzerland shows the casting right next to what seems to be the opening to the worm's tunnel. Many fossil trails of worm-like animals look similar to castings, but the fact that castings often show distinct ends serves to separate them. Castings are also unique in that they are a localized pile of randomly overlapping loops: a feeding or traveling organism's trails tend to be more orderly. Worm castings are some of the oldest animal fossils known.

Order:	n/a
Family:	n/a
Habitat:	Marine
Distribution:	Global
Time scale:	Proterozoic–Recent

Nemertean

Phylum Nemertea contains approximately 1000 known species of these widespread 'ribbon worms'. Typically long and thin, species vary in length from a few millimetres to 30m (98ft) (*Lineus longissimus*). Much of their time is spent beneath rocks or burrowing into soft sand on the water's edge. However, this is a remarkably adaptable phylum and worms are found in salt water and fresh water, in the shallows or ocean depths, and even on land – in fact, almost anywhere that is damp. Their diets, too, are incredibly varied. All of these brightly coloured, unsegmented worms are carnivores, but they feed on anything from other worms to molluscs. Some even live a parasitic existence inside the bodies of other animals such as crabs.

Order:	Various
Family:	Various
Habitat:	Shallow coastal waters
Distribution:	Global
Time scale:	Pre-Cambrian–Recent

Serpula sp.

Serpula possessed a chalky tubular skeleton ornamented with fine concentric rings that measured up to 10cm (4in) long, and was spirally or irregularly coiled. These tubes were normally cemented to hard, stationary objects. The tube skeleton's lid was chalky or horny. *Serpula* was found all over the world during the Silurian Era. In both the Jurassic and the Cretaceous, *Serpula socialis* was characterized by long, straight tubes, each with a diameter of 0.7mm (⅟₁₀₀in), and were connected to each other in bundles. A bundle of *Serpula* tubes could measure 6cm (2¼in) overall.

Class:	Polychaeta
Order:	Serpulimorpha
Habitat:	Marine
Distribution:	Global
Time scale:	Silurian–Recent

Didontogaster sp.

Didontogaster was a type of predatory worm that lived around 300 million years ago. It fed on sponges, corals and other simple animals that could not get away, crawling over them and biting off chunks at will. *Didontogaster* was related to modern-day bristleworms, marine cousins of the earthworms. Unlike most bristleworms, which live in burrows, it spent its time out in the open on the seabed, creeping around in the search for new prey. *Didontogaster* was a small creature, growing to just over 3cm (1¼in) long.

Class:	Polychaeta
Order:	Phyllodocemorpha
Habitat:	Seabed
Distribution:	Global
Time scale:	Late Carboniferous

Dictyodora isp.

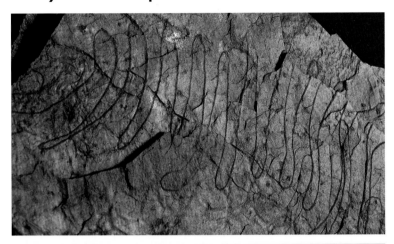

The graphoglyptids are a highly organized group of trace fossils that represent the activities of shallow mud-burrowers. *Dictyodora* is a graphoglyptid with regular parallel meanders that might have been made by a worm-like animal. These complex burrow systems were originally thought of as the trails of a methodical organism systematically covering a surface in the search for food. Traces preserving this type of activity are termed pascichnia ('grazing traces'). The prevailing thinking now places *Dictyodora* in the category of agrichnia ('gardening traces'). The meanders are presumed to be structured networks crafted to capture migrating meiofauna (small benthic invertebrates) or perhaps to culture bacteria. Certain organisms develop this type of 'farming' in nutrient-poor deep-sea environments.

Order:	n/a
Family:	n/a
Habitat:	Deep marine
Distribution:	Global
Time scale:	Pan-Phanerozoic

Hyolithes cecrops

Hyolithes cecrops was a member of the hyolithids, a puzzling group, global in distribution, that lived through most of the Palaeozoic. They were small bottom-dwelling animals, no more than 4cm (1½in) long, with conical calcareous shells that were triangular or elliptical in cross section. The shells sported an operculum (cover) over the wide end of the cone, which was flanked by two curved projections called helens. These may have acted as supports to keep the operculum open or as stabilizers in soft sediments or rough currents. Aside from their external anatomy, little else is known about them. None have yet been found preserving internal anatomy, which would help to explain their relationships, so their classification is still a mystery, although some scientists place them in the molluscs.

Class:	Hyolithomorpha
Order:	Hyolithida
Habitat:	Marine
Distribution:	Global
Time scale:	Middle Cambrian

Dentalium sp.

The scaphopods, or tusk shells, are a very long-lived group of marine molluscs that have undergone very little apparent change in hundreds of millions of years, showing very little diversity of form within the group. Their gently curved conical shells, sometimes decorated with longitudinal or concentric ridges, taper gradually and end in a small hole, into which water was drawn and expelled from the soft anatomy for purposes of eating and respiration. The animal sits in the mud with its narrow end up and extracts organic material from the sediments with minute tentacles called captacula. The shells of *Dentalium* were painstakingly dredged from the sea bottom by natives of the Pacific Northwest and used as a form of money or sacred object called wampum.

Class:	Scaphopoda
Family:	Dentaliidae
Habitat:	Marine muds
Distribution:	Global
Time scale:	Ordovician–Recent

Chitonid

Chitons are among the most ancient animals alive today. The earliest chitons appeared in the Cambrian Period more than 520 million years ago. They are molluscs related to slugs and snails and, like them, have a single muscular foot. Unlike snails and slugs, they have a shell made up of eight overlapping plates. Because they have existed for so long, chiton fossils appear in rocks from many different stages in the Earth's history. They are easy to recognize because their shells look unlike those of any other group of animals. Nowadays, most Chitons are found in coastal waters, where they feed on algae growing on rocks. There are a few species that live in the deep sea and feed on bacteria.

Order:	Neoloricata
Family:	Chitonidae
Habitat:	Marine
Distribution:	Global
Time scale:	Cambrian–Recent

Gryphaea sp.

The coiled shell of *Gryphaea arcuata* has the popular name of the Devil's Toenail. Its fossils have been found all over the world. This bivalve had a right valve that was concave and provided a lid for the much larger left valve. The left valve featured a narrow beak that was rolled inwards. The surfaces of both right and left valves were covered in several coarse ridges. *Gryphaea arcuata* had one circular muscle scar and the hinge in adults was toothless. Typically 7cm (2¾in) in length, *Gryphaea arcuata* had a heavy calcite shell and was adapted for living in the soft sediment of its muddy seabed habitat. Young *Gryphea* attached themselves to hard surfaces by their tips.

Class:	Bivalvia
Order:	Pterioida
Habitat:	Marine
Distribution:	Global
Time scale:	Late Triassic–Late Jurassic

Spondylus sp.

Also known as the thorny or spiny oyster, *Spondylus* is the only genus in the family Spondylidae. It is closely related to the scallops, which also have wings on the sides of their beak as seen on this specimen. Modern *Spondylus* also has eyes like the scallops, and it can be assumed that the ancient spondylids did, too. They differed, however, from the scallops in their life mode: scallops locomote through the water by using jet propulsion, but *Spondylus* led a sessile life, like the oysters, attached to hard surfaces. One Cretaceous species, *Spondylus gregalis*, is often found fused to the inside of larger, heavy-shelled oyster shells. The hinge of these bivalves was a ball and socket joint rather than the toothed hinge more common in bivalves.

Order:	Pterioida
Family:	Spondylidae
Habitat:	Marine
Distribution:	Global
Time scale:	Middle Jurassic–Recent

Teredo sp.

Shipworms are the bane of wooden-hulled ships, and have been for centuries. They drill into wood, where they live out their lives, causing severe structural damage. Misnamed 'worms', shipworms are actually bivalves of the genus *Teredo* whose paired shells have all but disappeared as they evolved a somewhat worm-like existence. But shipworms need no ships, and as long as wood has been making its way into the sea, there have been animals like *Teredo* waiting to exploit it as a floating place to live. This is a piece of wood from Late Cretaceous Cannonball Formation of North Dakota and clearly shows the crowded tunnels of *Teredo* all reaching to the outer surface of the wood, where they would filter their food from the water.

Class:	Bivalvia
Family:	Teredinidae
Habitat:	Marine (in wood)
Distribution:	Global
Time scale:	Eocene–Recent

Trigonia sp.

The coelacanth is famous for being an animal initially presumed to be long extinct before being found alive and well in the present. It was thought to have gone extinct at the end of the Cretaceous. But the bivalve *Trigonia* is one of many that fall into this category. Long before the 1938 discovery of living coelacanths off South Africa, *Neotrigonia* was found in 1802 in Australia, proving the tenacity of this group of bivalves. They are burrowing molluscs, and it has been shown that the knobs and grooves on the shells, which vary greatly among species, aid them in burrowing into the sea bed. The animal sits with just enough of itself exposed to draw water in for oxygen and food, and to expel waste.

Order:	Trigonioida
Family:	Trigoniidae
Habitat:	Sea bed
Distribution:	Global
Time scale:	Triassic–Recent

Ostrea sp.

Oysters are bivalves that differ from most bivalves by having shells completely comprised of calcite but with internal muscle scars of aragonitic composition. This single muscle scar can be seen in the two *Ostrea* shells on the left. Another thing common to oysters is their generally attached life habit: their shells fuse onto rocks, wood, or other hard surfaces like shells, as seen here where three oysters have attached themselves to the shell of a dead scallop. This habit of attaching is especially true of young oysters – adults are often large enough to live free on the bottom. Attaching also leads to the irregular shells of oysters: oysters found in rocky vs. muddy substrates can differ markedly, even if they are the exact same species.

Order:	Pterioida
Family:	Ostreidae
Habitat:	Marine
Distribution:	Global
Time scale:	Triassic–Recent

Pecten sp.

Most bivalves live their adult life embedded in mud or attached to a surface of some kind. The pectens, or scallops, have a more active life. Strong muscles are used to pump water forcefully out of the shell, which propels the pectin through the water out of harm's way or to a better location of its choosing. Also odd for a bivalve, the pectens have many eyes. Although they have poor eyesight, they can detect changes in light, signalling danger from predators or slumping sediments. The shell shown here is encrusted with the calcareous tubes of serpulid worms, indicating that this shell had lain on the seafloor for some time before burial. This allowed the serpulids, which seek stationary substrates, to adhere to the shell.

Order:	Ostreoida
Family:	Pectinidae
Habitat:	Marine
Distribution:	Global
Time scale:	Middle Devonian Period–Recent

Venus sp.

The round shell of the bivalve *Venus*, which measured around 3.6cm (1½in) long, had a thick wall and valves that were mirror images of each other. This is one way to separate clams from brachiopods whose line of symmetry tends to divide each shell rather than run along the junction of the two. There was an outer ligament positioned behind the umbones (where the development of the shell began). The external surface of *Venus*' shell was decorated with thin concentric lines. This bivalve lived mainly in North America, Europe and Asia, and its fossils have been found in Jurassic sediments. Another so-called 'living fossil', *Venus* has survived to the present.

Class:	Bivalvia
Family:	Veneridae
Habitat:	Marine
Distribution:	Global
Time scale:	Oligcene–Recent

Inoceramus sp.

Inoceramus was a common bivalve whose fossils are found all over the world, especially in North America's Western Interior Seaway, which bisected the continent in the Cretaceous. Some members of its family grew to enormous size, becoming huge disks several feet across. This was presumably an adaptation to spread their weight wide to avoid sinking into the soft chalky ooze they inhabited. One inoceramid species, *Platyceramus platinus*, has been found with its delicate gills wonderfully preserved within. But what is even more amazing, is that within those gills are the skeletons of dozens of tiny fish that lived within the living bivalve. It is presumed that these fish sought shelter in the huge molluscs, which were, seemingly, the only places to hide in their environment.

Order:	Pterioida
Family:	Inoceramidae
Habitat:	Sea bed
Distribution:	Global
Time scale:	Cretaceous–Miocene

Volviceramus sp. (pearls)

Contrary to popular belief, pearls are probably not commonly formed by a grain of sand irritating a bivalve. The concentric layers of nacre, or mother-of-pearl, that form a pearl actually accrete around something like a tiny shrimp or other organic debris. The pearls seen here were found in the Late Cretaceous Smoky Hill Chalk of Kansas and were formed by the giant bivalve *Volviceramus grandis*. The largest pearl pictured, 2cm (¾ in) wide, is a blister pearl, as is the one on the right. Blister pearls form when the object of irritation gets sealed against the mollusc's shell by the enclosing nacre. Bivalves are not the only pearl-forming molluscs: gastropods, nautiloids and even ammonoids have been found to produce them as well.

Order:	Pterioida (*Volviceramus*)
Family:	Inoceramidae (*Volviceramus*)
Habitat:	Marine & Freshwater (pearls)
Distribution:	Global (pearls); Texas and Kansas (*Volviceramus*)
Time scale:	Pan-Phanerozoic, the oldest-known are upper Triassic (pearls)

Arctostrea (Rastellum) sp.

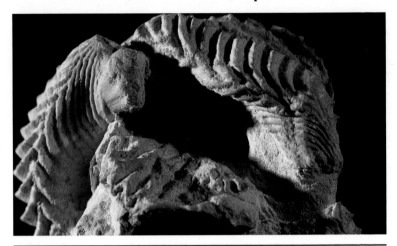

*A*rctostrea, also known as *Rastellum*, had a shell characterized by deep vertical pleats along its entire length. The shell, which was on average 10cm (4in) long, comprised two valves. The pleats were so arranged that it would have been difficult, if not impossible, for a predator to find the division between the valves in order to prise them open. It was particularly vital for *Arctostrea* to have this defence as it was a stationary creature, spending its life anchored to hard surfaces in the warm waters it inhabited. *Arctostrea*, like most oysters, have valves that differ from each other markedly. This contrasts with most ofther bivalves, whose two shells are mirror images of each other.

Class:	Bivalvia
Order:	Pterioida
Habitat:	Ocean floor
Distribution:	Europe, USA
Time scale:	Middle Jurassic–Late Cretaceous

Arctostrea carinatum

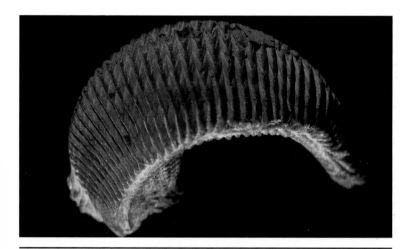

Arctostrea carinatum was an ancient relative of modern oysters, although it looked very different from most oysters that are around today. In North America, it has been given the nickname 'denture clam' because of its weird shape – many people think that it looked like a set of false teeth. *Arctostrea carinatum* was a filter feeder that sifted its food from the water around it. Like its modern relatives, it sucked water in through one tube and squirted it out from another. In between, the water passed over the sieve-like gills, which removed tiny particles of food and absorbed oxygen. This little shellfish lived on the seabed around 100 million years ago. It grew to about 5–6cm (2–2½in) long.

Class:	Bivalvia
Order:	Pterioida
Habitat:	Marine
Distribution:	Global
Time scale:	Cretaceous

Claraia sp.

A t first sight, *Claraia* looks pretty unimpressive. It is a small bivalve mollusc – a shellfish not unlike a modern-day mussel or clam. However, *Claraia* did something that few other creatures were able to manage. It survived through the mass extinction event that happened 245 million years ago. Scientists are unsure exactly what caused this event, although it might have been a collision between the earth and a massive rock from space. What is known for certain is that this event wiped out around three-quarters of all of the species on Earth. It separates the Permian Period from the Triassic Period. With most of its competition killed off, *Claraia* became extremely successful after the extinction event and spread to areas of the ocean floor where it had not existed before.

Order:	Pterioida
Family:	Aviculopectinidae
Habitat:	Ocean floor
Distribution:	Global
Time scale:	Permian–Triassic

Pholad (borings)

Many bivalves bore into hard substrates for the added protection they afford. This burrowing occurs in one of two ways: chemical or mechanical. Chemical borers secrete acids that slowly etch the rocks in which the clam intends to live. Mechanical borers have a drill-like end that is ground against wood or rock until a hole is made for the clam to live in. As the clam grows, it continues drilling, and flushes out the loosened particles from the hole. These Cretaceous borings from New Jersey show the pear-shaped excavations of mechanical borers that prevent the clam from being pulled from its hole. These are some of the few fossils that originate in rock, as opposed to being incorporated into rock; and if found out of context, the rock cannot be dated.

Order:	Myoida
Family:	Pholadidae
Habitat:	Marine rocks
Distribution:	Global
Time scale:	Jurassic–Recent

Exogyra sp. and *Gervilliopsis ensiformis* (xenomorph)

E*xogyra* was a type of oyster and, like most in its group, its young needed a substrate to fuse onto early in life. Any hard, stationary surface would do, such as wood, bones, or rock. In the case of this Late Cretaceous specimen from New Jersey, its spat landed on the shell of *Gervilliopsis ensiformis*, a razor clam. The flattened part on the right clearly shows that coils of the baby oyster. The flat area, which is a mold of the *Gervilliopsis*, is called a xenomorph: a facsimile of one organism retained in the surface of another. The razor clam has all but disintegrated, leaving just a hint of its original pearly luster. Some fragile species are saved in this way when the sediments themselves had little chance of preserving them.

Order:	Pterioida (*Exogyra* and *Gervilliopsis*)
Family:	Gryphaeidae (*Exogyra*); Bakevelliidae (*Gervilliopsis*)
Habitat:	Marine (both taxa)
Distribution:	Global (*Exogyra*); North America (*Gervilliopsis*)
Time scale:	Jurassic–Cretaceous (*Exogyra*); Cretaceous (*Gervilliopsis*)

Diceras angulatum

Diceras was one of the first of a bizarre branch of the bivalve tree called rudists. Rudist bivalves grew heavy, often contorted shells that in some forms resemble the heavy solitary corals of the Palaeozoic. These types apparently led a similar life to corals, forming huge reefs. In species like *Diceras*, the umbones (beaks) twisted into heavy corkscrews that helped the animal secure itself in variable substrates. But it was not actually the beaks that were twisting: it was the junction of the two shells where additional shell material was added during growth. The two shells are not mirror images of one another as in most other bivalves. The attached valve (shell) is usually the larger one.

Order:	Hippuritoida
Family:	Diceratidae
Habitat:	Marine
Distribution:	Europe
Time scale:	Late Jurassic Period

Hippurites sp.

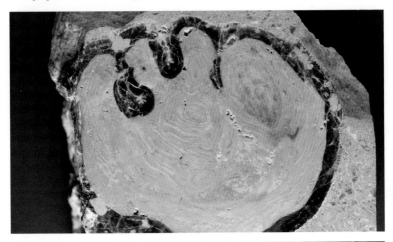

*H*ippurites was a part of a specialized group known as rudists. This bivalve had dissimilar valves – one shaped like a deep cone, the other a flat valve that acted as a 'lid'. The outside of both valves was covered by a fluted surface. Within the 'chamber', the lid had two long teeth growing from its inner surface. Also, muscles were attached to two bosses. The inner walls of the cone-shaped valve were folded into pillars, together with a narrow tooth and sockets designed to hold the tooth in the lid. Some lived solitary lives, but others preferred to group together and formed reefs.

Class:	Bivalvia
Order:	Hippuritoida
Habitat:	Marine reefs
Distribution:	Global
Time scale:	Late Cretaceous

Actaeonella sp.

Actaeonella is a genus of fossil snail. However to the untrained eye, that might be hard to discern from this specimen. Their random orientation and sectioning could confuse one into thinking they might be petrified eyes, nuts, eggs or maybe pine cones. The history of palaeontology is replete with folkloric descriptions of fossil origins and their uses. *Actaeonella* fossils were once called 'wirfelstein' or 'wirbelstein', meaning 'spiral stones' in German, and were put into the drinking troughs of sheep to prevent them from getting a certain worm-borne disease that caused the sick sheep to rotate. Presumably, the behaviour of worms and stricken sheep, plus the spiralling shells of *Actaeonella*, were all seen as related, which was common in folk remedies a few centuries ago.

Class:	Gastropoda
Order:	Cephalaspida
Habitat:	Marine
Distribution:	Europe, Africa & the Americas
Time scale:	Cretaceous

Bellerophon sp.

*B*ellerophon, named after the hero of Greek mythology who killed the Chimera, was a genus of gastropods of unusual pattern. It is characterized by a globose, involute (with overlapping coils), planispiral (symmetrically coiled) shell, much like certain ammonites but lacking the septa that all shelled cephalopods have. Unlike the specimens shown here, they often had a groove or ridge running along the midline of the shell, the function of which remains unclear. The genus tends to be regarded as a primitive snail, but a minority of scientists sees them as a separate group of molluscs that acquired a spiral shell independent of the gastropod lineage. Unfortunately, many of the features that could be used to resolve this issue are soft tissue anatomy, which decays easily and is rarely preserved.

Order:	Bellerophontida
Family:	Bellerophontidae
Habitat:	Marine
Distribution:	Global
Time scale:	Ordovician–Early Triassic

Xenophora sp.

Xenophora's shell was fairly short and conical. Its suture (the line where the whorls of gastropod shells connect) was not very deep. The shell had a slanting profile, interrupted at intervals. The underside was flat. The narrow aperture was covered by the thin edge of the shell. *Xenophora* lived, and still lives right across the world feeding on microrganisms that populated the muddy bottom of the seas. At some time, *Xenophora* developed the ability to collect a shell or other suitable fragment of material on the sea bottom and cement itself to its surface. This was done for the purpose of camouflage, so that *Xenophora* could escape the attentions of potential predators as they passed by in search of food.

Class:	Gastropoda
Order:	Sorbeoconcha
Habitat:	Marine
Distribution:	Global
Time scale:	Cretaceous–Recent

Xenophora sp.

The medium-sized mollusc *Xenophora* was cone-shaped with a flattened base and measured about 8cm (3¼in) in height. *Xenophora* fossils have been found all over the world. Also known as the 'carrier shell', *Xenophora* adorned itself with all manner of debris – from shells, shell fragments, rocks, coral and (in modern species) glass – for camouflage. These bits were incorporated into the growing shell throughout its life, and different species tend to be consistent in the objects they collect.

Class:	Gastropoda
Order:	Sorbeoconcha
Habitat:	Marine
Distribution:	Global
Time scale:	Cretaceous–Recent

Euomphalus sp.

Gastropods are the only known molluscs that successfully made the transition between aquatic and terrestrial environments. Accordingly, the terrestrial forms underwent significant changes in order to live exclusively on land. Aside from the obviously necessary changes in breathing anatomy, land snails have eyes on stalks and a pair of feelers. Fresh water snails also have feelers, but of different arrangement, and often more than one pair and their eyes are unstalked. Marine snails have feelers but, like other aquatic gastropods, have no eyestalks. Although the soft anatomy of the marine snail *Euomphalus*, seen here, is unknown, it can fairly safely be assumed that it had the stalkless eyes of its modern marine relatives. This interpretation is based on the available evidence, but is not verifiable.

Class:	Gastropoda
Order:	Euomphalina
Habitat:	Marine
Distribution:	Global
Time scale:	Silurian–Permian

Gastropoda (opalized)

Once the remains of organisms are buried in the ground, geological processes act upon them. They get no preferential treatment – whatever happens to rocks and other sediments will happen to organismal remains. The vast copal fields of Australia formed in palaeontologically rich sediments, and the fossils found there are often preserved in precious opal. This small Cretaceous snail shows the green fire of the opal that has replaced it. Petrified wood, dinosaur teeth and even whole skeletons of marine reptiles have been unearthed ablaze with the fire of opal. Unfortunately, museums can rarely afford the extremely high prices these specimens demand, and the opal miners wind up breaking up the fossils to be sold as much more lucrative gem opal.

Order:	Gastropoda
Family:	n/a
Habitat:	Marine
Distribution:	Opalized gastropods are found in Australia
Time scale:	Cretaceous

Poleumita discorus

*P*oleumita discorus was a coiled gastropod measuring between 5cm (2in) and 9cm (3½in) wide. Its shell was ornamented by moderately raised spines on its shoulder, the part of the shell where it turns inwards towards the suture (line where the whorls meet). Fine lamellae also decorated the *Poleumita's* shell. For a gastropod, *Poleumita's* appearance was unusually angular. *Poleumita* fossils have been found in North America and in Europe, where it proved prolific in the county of Shropshire in western England. They display a wide range of types – some with flattened shells, others with cone-shaped shells. Most, though, have delicate ribbing and smooth undersides.

Class:	Gastropoda
Order:	Euomphalina
Habitat:	Marine
Distribution:	Europe & North America
Time scale:	Silurian

Strombus sp.

For a gastropod, *Strombus* could grow to significant size. *Strombus'* shell had thick walls and its coils were wound helix-like in the form of a low three-dimensional spire. The last of its convex whorls was large and acted as the base of the shell. *Strombus'* aperture was usually oval-shaped, but was sometimes oblong. Although sometimes smooth, the outside of *Strombus'* shell was more often decorated with spines on the edge of the whorls. Fossils of the genus *Strombus* have been found over a wide area – in Africa, Europe and Asia. *Strombus* survives to this day, but lives in more confined circumstances, in the Indian Ocean, Pacific Ocean, western India and Japan.

Class:	Gastropoda
Family:	Strombidae
Habitat:	Marine
Distribution:	Global, low-latitudes
Time scale:	Cretaceous–Recent

Cerithium sp.

*C*erithium shells usually grew a large number of thick whorls that formed a high tapering turret. The shell walls were themselves thick and were ornamented by tubercles which formed a pattern of spiralling bumps. *Cerithium* had a broad oval-shaped sloping aperture and a short, deep tubular canal for taking in water. The canal had a distinct curve. The inside of *Cerithium's* outer lip often exhibits a wrinkled surface. *Cerithium* survives to this day and is now found in all the warmer areas of the world. *Cerithium* could grow to a height of 13cm (5in).

Class:	Gastropoda
Family:	Cerithiidae
Habitat:	Marine
Distribution:	Global, mid-latitudes
Time scale:	Jurassic–Recent

Cypraea sp.

*C*ypraea is the scientific name for the molluscs most people know as cowries. These marine creatures are snails and, like all gastropods, have a single large foot which they use for getting around. Most cowries today feed on algae, although a few eat coral polyps. It seems very likely that their ancestors, which we find as fossils, lived in the same way. Although *Cypraea* fossils are quite beautiful in shape, one thing that is not preserved is their colour. Most living cowries are known to be very colourful creatures with shiny shells often decorated with beautiful patterns. When they sense danger, they pull their fleshy bodies into their shells for protection, like other snails.

Order:	Sorbeoconcha
Family:	Cypraeidae
Habitat:	Marine
Distribution:	Global
Time scale:	Cretaceous–Recent

Natica sp.

Natica had a smooth round shell with a large body whorl, although its spire (set of whorls) was quite small. Informally known as the moon shell, *Natica's* operculum, which was attached to its foot, was there to close the shell and made it a tight fit. *Natica* fossils have been found all over the world in marine waters as well as brackish waters and at a variety of sea and ocean depths. This gastropod fed itself by drilling holes in the shells of other molluscs and sucking out whatever it found inside. Although rarely spotlighted as such, *Natica* is yet another 'living fossil', and boasts a history of more than 390 million years.

Class:	Gastropoda
Order:	Neotaenioglossa
Habitat:	Marine
Distribution:	Global
Time scale:	Middle Devonian–Recent

Turritella sp. (steinkern)

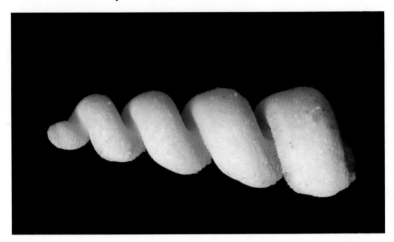

When people think of fossils, they tend to think of hard parts like bones and shells and petrified wood. It is true that remains like these have the best chances of being preserved, but sometimes only an impression in the stony sediments of the surface of a hard part exists. These are called moulds. And if the original material totally disappears, leaving only a mould, a natural cast might form which replicates the surface of the original. Here we have an internal mould of a snail, probably of the genus *Turritella*. Sediments filled the living chamber of the dead snail and turned to stone, creating a steinkern – in German, a 'stone core'. Many deposits that are hostile to the preservation of shelly material may have abundant steinkerns.

Order:	Neotaenioglossa
Family:	Turritellidae
Habitat:	Marine
Distribution:	Global
Time scale:	Late Triassic–Recent

Helix sp.

The land snail *Helix* belongs to a genus whose fossils have been dated from the Oligocene, some 33 million years ago. *Helix* once lived all over Europe but, having survived until today, is now confined to the the area around the Mediterranean and the Black Seas. *Helix's* 2.6cm (1in) shell had thin walls and was normally spherical. There was a low blunt spire but few whorls. The body whorl, however, was large. The shape of *Helix's* aperture can vary, which limits the work required to identify a number of species. *Helix's* aperture, for instance, can be oval or semi-circular. Today *Helix* is a popular snail for escargot.

Class:	Gastropoda
Order:	Styilommatophora
Habitat:	Terrestrial
Distribution:	Europe
Time scale:	Oligocene–Recent

Murexsul sp.

The gastropod *Murexsul* was on average 9cm (3½in) long, with a medium-sized shell that featured sharp outstanding ribs. *Murexsul's* whorls were sharp and its suture (the line on gastropods where its whorls connect) was deep. The walls of *Murexsul's* whorls were convex and the whorls themselves were large for a gastropod of its size. The general effect resembled a spiral turret with an exterior formed from fluted spines. The aperture (the opening at the shell margin) was surrounded by the last rib and was smooth and glassy. The base of the aperture suddenly narrowed into a long tube-shaped canal, which *Murexsul* used to take in water. *Murexsul*, which was a carnivore, lived in fairly deep waters, where it consumed other animals.

Class:	Gastropoda
Order:	Neogastropoda
Habitat:	Marine
Distribution:	Pacific
Time scale:	Pliocene–Recent

Tentaculites sp.

*T*entaculites is an enigmatic genus of invertebrates with ribbed conical shells. They are very common in some sediments, where their tiny 5–20mm (⅕–⅘in) shells comprise the entire surface of certain layers. But despite their omnipresence, they still defy classification. This is because they display few of the features necessary to place them firmly in known categories, especially since, for many organisms, it is the soft parts that define them – only the shells of *Tentaculites* have been found. Over the years, they have been considered worms, cephalopods, and, most commonly, members of a group of gastropods called pteropods. It is altogether possible that they represent a new group of previously known animals, something altogether new, or even mere pieces of another sort of animal, known or unknown.

Order:	Tentaculitida
Family:	n/a
Habitat:	Inland seas
Distribution:	Global
Time scale:	Early Ordovician–Late Permian

Aturia sp.

Aturia was the most common member of the Order Nautilida. During the Tertiary, it spread world-wide. *Aturia* came in various sizes with a conical shell – spirally coiled to form a disc, completely smooth on the outside – which could grow from only 2cm (¾in) to 20cm (7¾in) in diameter. The sutures (the lines where the chambers met) bore a wide, angular 'saddle' on its venter (outer edge). The sides of *Aturia's* sutures carried a narrow lobe and another saddle, this time a rounded one. Around 50 species of *Aturia* have been recorded over a wide area – in Europe, Africa, Asia, North and South America and Australia.

Class:	Cephalopoda
Order:	Nautilida
Habitat:	Maritime
Distribution:	Global
Time scale:	Palaeocene–Miocene

Lituites sp.

*L*ituites had an unusual shell. Only the first whorls were coiled in a flat spiral. The rest of it was a straight cone, making the whole shell resemble an elephant's trunk. The majority of shelled cephalopods throughout Earth history have been fully coiled, allowing us to compare them with today's coil-shelled cephalopod *Nautilus*. But oddities like *Lituites* leave us wondering about their life habits. Early members of the genus *Litruites* included some of the first cephalopods – molluscs of a class that included squid and octopus. *Lituites* was a small nautiloid – its shell measured around 12cm (4¾in) long.

Class:	Cephalopoda
Order:	Orthocerida
Habitat:	Marine
Distribution:	Global
Time scale:	Ordovician

Cenoceras sp.

Cenoceras' shell curled in a spiral and had a narrow umbilicus, the point around which the shell coiled. As the whorls progressed, their height increased rapidly. The venter (outer edge) was round and broad. There were no complex folds along the suture lines (where the chambers met). Inside the creature's venter, there was a long, thin siphuncle (the tube-shaped extension of the main body that passed through all the chambers of the shell). The siphuncle was used to regulate the gas in the animals chambers to adjust its buoyancy. All chamber-shelled cephalopods used this system of gas buoyancy.

Class:	Cephalopoda
Order:	Nautilida
Habitat:	Marine
Distribution:	Europe & North America's northern latitudes
Time scale:	Jurassic

Nautilus sp.

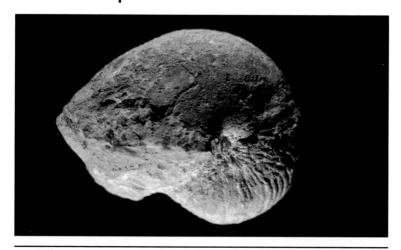

This animal is a relative of the ammonites and belemnites, but unlike them it survived the mass extinction event at the end of the Cretaceous Period and is still around today. The modern nautilus looks almost identical to its fossils from rocks laid down 420 million years ago. This makes it one of the most primitive animals alive on Earth today. Like the now long-extinct ammonites, *Nautilus* is a type of mollusc with tentacles and a hard outer shell. It uses its tentacles to catch fish and other prey in open water. Modern-day nautiluses have quite colourful shells, with streaks of red splashed over a white background. It seems more than likely that they were colourful in ancient times as well.

Class:	Cephalopoda
Order:	Nautilida
Habitat:	Open ocean
Distribution:	Global
Time scale:	Ordovician–Recent

Orthoceras sp.

Orthoceras belonged to a primitive family of marine creatures with shells. *Orthoceras'* shell, a long tapering cone, was composed of several chambers formed close to one another, with a siphuncle running through them. *Orthoceras* was around 15cm (6in) in length. It was a strong swimmer, moving rapidly by squirting water out of its body chamber. When swimming, deposits of calcium carbonate in the central area of *Orthoceras'* shell acted as a counterweight, allowing the shell to lie lengthwise in the water. *Orthoceras* fossils have been found in Europe. Nautiloids as such left their fossils all over the world and were particularly prolific during the Palaeozoic Era 400 million years ago.

Class:	Cephalopod
Order:	Orthocerida
Habitat:	Marine
Distribution:	Global
Time scale:	Early Ordovician–Triassic

Gastrioceras sp.

*G*astrioceras was related to *Goniatites*, featuring the characteristic flattened shell of the Goniatitida order of molluscs. *Gastrioceras* shell was round with flat, broad whorls that overlapped slightly and were rounded at the edges. The height of the whorls was less than their breadth. *Gastrioceras* was usually around 2.5cm. (1in.) in diameter. It appears to have been a slow swimmer, but considering its habitat, there was no great need for speed. *Gastrioceras* lived in the shallow waters at the coastal edges. The two specimens here are natural casts in stone – the original shell has long since disintegrated.

Class:	Cephalopoda
Order:	Goniatitida
Habitat:	Marine
Distribution:	England, USA
Time scale:	Carboniferous

Laevaptychus sp. (jaws)

All cephalopods have beaks at the centre of their array of arms. Aptychi (singular aptychus) is the name given to the paired lower jaws of ammonites, a cephalopod group that became extinct 65 million years ago. Since they are usually found separated from the ammonite shells they were associated with in life, aptychi were given their own names, like *Laevaptychus* shown here. Although some early scientists correctly deduced what they were, it was not until they were found in place in the shell that it could be said with confidence that they were, indeed, jaws. They are less frequently preserved, partly because they were composed of chitin and calcite, whereas the shells of ammonites were composed of aragonite: the composition of aptychi decayed much more readily than argonite.

Class:	Cephalopoda
Order:	Ammonitida
Habitat:	Marine
Distribution:	Europe
Time scale:	Late Jurassic

Leioceras opalinum

*L**eioceras opalinum* was an actively mobile nektonic carnivore – it swam around in the water column, hunting smaller prey. This ammonite had a geologically very short existence, which allows it to function as a place marker in the rock record. If you find a fossil of *Leioceras opalinum*, you can be reasonably sure you are in the Middle Jurassic; and if you find it along with fossils of *Leioceras lineatum*, the resolution gets even tighter, putting you near the base of the Aalenian Stage of the Middle Jurassic some 180 million years ago. In this specimen, you can see the wavy lines of the external shell ornamentation. These should not be confused with the species diagnostic sutures that lie perpendicular to this surface just under the shell.

Class	Cephalopoda
Order:	Ammonitida
Habitat:	Open water
Distribution:	Europe
Time scale:	Middle Jurassic

Dactylioceras sp.

Dactylioceras, another ammonite, had an evolute (loosely coiled) shell. There were many closely-spaced ribs that continued over its rounded venter. In some fossils, the outer whorl of *Dactylioceras'* shell ended in a feature shaped like a spatula extending from the edge of its aperture (the opening to the animal's living chamber). Ammonites, like *Dactylioceras,* were once called 'snake stones' because they were thought to be snakes petrified by a famous religious figure. Snakes heads were often carved onto the ammonities in medieval times, for sale to pilgrims at religious sites. *Dactylioceras* measured about 7cm (2¾in) in diameter and was a slow swimmer.

Class:	Cephalopoda
Order:	Ammonitida
Habitat:	Marine
Distribution:	Northern latitudes
Time scale:	Early Jurassic

Baculite sp. (pyritized)

Straight shelled ammonites of the genus *Baculites* are extremely common fossils in some strata of the western United States. This one, from the Cretaceous Pierre Shale of South Dakota, has been beautifully preserved in pyrite. The chambers of the ammonite have been filled with pyrite and the outer layers of shell have worn away, exposing the wavy suture lines embedded in the lustrous mineral. Ammonite sutures are very useful in distinguishing species, and often the smallest bit of preserved suture can tell an expert what species it is. Where this specimen comes from, the native tribes also had a use for the chamber fillings sometimes found loose on the surface: they called them buffalo-calling stones, for their slight resemblance to bison, and used the fossil's magic to summon bison.

Class:	Cephalopoda
Order:	Ammonitida
Habitat:	Marine
Distribution:	Global
Time scale:	Late Cretaceous

Ammonite

The ammonites were an incredibly diverse group of cephalopods that populated the seas for many millions of years. They secreted external calcareous shells and, just like many of their modern molluscan relatives, these shells could exhibit nacre, otherwise known as mother-of-pearl. The microstructure of nacre is such that light hitting it can throw off a changing rainbow of colour. Unlike colours that arise from pigments, which are chemicals that can break down over time, the structure of nacre can often survive millions of years in the fossil record and still have the beautiful play of colour of the original shell. Such is the case with this ammonite fossil. The animal, however, may not have looked like this in life since nacre is often hidden within the shell or under a differently coloured exterior.

Class:	Cephalopoda
Order:	Ammonitida
Habitat:	Marine
Distribution:	Global
Time scale:	Permian–Cretaceous

Ammonite

Ammonites are most closely related to the octopus, but the animal that comes to mind more often when one pictures them is the nautilus. *Nautilus* is the only living taxon of cephalopod with a coiled external shell throughout its life. The shell is divided into chambers that were filled with gas which was governed by a thin strand of tissue called the siphuncle. The gas in these chambers was adjusted by the nautilus to alter its buoyancy and thus its position in the water column. Ammonites had this same buoyancy-regulating apparatus. This fossil shows a split ammonite in which the chambers were not filled with sediment after death or minerals after burial. Whereas *Nautilus* has smooth cupped septa between chambers, ammonites, like this one, had complex wavy septa.

Class:	Cephalopoda
Order:	Ammonitida
Habitat:	Marine
Distribution:	Global
Time scale:	Permian–Cretaceous

Ammonite

This rare ammonite fossil shows the placement of its lower jaws, known as aptychi. The line separating the two halves is the animal's midline, which shows that the jaws have rotated about 90° to the left within the living chamber of the shell, seen here showing its left side. Just like in their modern analogue, the nautilus, they had a chambered shell with gas-filled chambers that did not extend to the opening of the shell. The large living chamber spanned the distance between the shell's aperture and the first septum and is where most of the animal's flesh was housed. Extending from the opening in the shell were numerous arms and the ammonites eyes, but the jaws were likely to be sequestered out of sight in a mass of flesh at the centre of the arms.

Class:	Cephalopoda
Order:	Ammonitida
Habitat:	Marine
Distribution:	Global
Time scale:	Permian–Cretaceous

Ammonite

In books and museum exhibits, we are accustomed to seeing the best and most complete palaeontological specimens. But these are by far the overwhelming minority of fossils unearthed. That is not to say that worn, fragmentary, or broken specimens are not beautiful, rare or informative. This odd chunk is a small water-worn fragment of an ammonite. The pair of complex sutures immediately say to the trained eye, 'ammonite'. This specimen was found on Banks Island in the Canadian Arctic, now a cold, treeless tundra. But this small bit of ammonite comes from an animal no less than 65 million years old, which lived in a much warmer place and in the sea. Even the smallest recognizable piece of a fossil can speak volumes to those who can hear.

Class:	Cephalopoda
Order:	Ammonitida
Habitat:	Marine
Distribution:	Global
Time scale:	Permian–Cretaceous

Crioceratites sp.

C*rioceratites* fossils have been found all over the world. The predatory *Crioceratites* hunted for food in fairly deep seas. On average, its diameter measured about 10cm (4in). The *Crioceratites* was coiled very loosely, with the outer coil almost three times thicker than the inner coil. The large outer coil carried whorls, while the much smaller inner coils had ribs with tubercles (small rounded projections). There were two or three finer ribs lacking tubercules between those that carried tubercles. The loosely coiled shell in *Crioceratites* has also been found in various nautiloids which lived between about 400 million and 245 million years ago, and in their immediate successors, the ammonites of the Mesozoic.

Class:	Cephalopoda
Order:	Ammonitida
Habitat:	Marine
Distribution:	North America, Europe, Africa, Asia
Time scale:	Cretaceous

Pleuroceras sp.

The ammonite *Pleuroceras* had a particularly striking evolute (loosely coiled) shell with a wide umbilicus. The whorls of *Pleuroceras* were nearly square when viewed in cross sections and the edges are only slightly convex. Thick, well-spaced radial ribs decorate the surface of this ammonite and finish with tubercles (bumps) at the edge of the umbilicus and on the outer edge. On the shell edges or venter, a braid-like keel can be seen. During the Lower Jurassic Era, *Pleuroceras* inhabited Europe and North Africa. One example of *Pleuroceras*, *Pleuroceras spinatum*, had a shell measuring 5.5cm (2¼in).

Class:	Cephalopoda
Order:	Ammonitida
Habitat:	Marine
Distribution:	Europe & Canada
Time scale:	Jurassic

Pleuroceras sp.

This impressive accumulation of fossil ammonite shells is known as a thanatocoenosis, or death assemblage. Since these animals are swimming predators and they are all lying on their sides with random orientation, it can be assumed that they all died and came to rest in a pile on the sea floor. (A life assemblage is a group of dead organisms preserved as they were in life, like a reef community). This may be the result of one event that killed them all at once, or it might represent a gradual accumulation of shells over time. But the fact that no other species are seen with the ammonites and that there is little sediment separating them suggests that this was a school of ammonites that met their end together.

Class:	Cephalopoda
Order:	Ammonitida
Habitat:	Marine
Distribution:	Europe & Canada
Time scale:	Jurassic

Phylloceras sp.

*P*hylloceras was a small- to medium-sized ammonite with a tightly coiled, compressed shell. Its average diameter was around 10cm (4 in) and its suture (the line where the chambers meet) highlighted here with pigment was noticeably frilled. Other ornamentation included growth lines or intermittent grooves, but some *Phylloceras'* shells were smooth. The aperture (the opening to the living chamber) curved gently. *Phylloceras* had a more or less streamlined shell and a rounded outer edge, so that it was able to swim at moderate speeds, using jet propulsion. *Phylloceras'* fossils have been found all around the world. However, they died out in the great extinction event that destroyed several other marine groups, such as belemnites and, most famously, the non-avian dinosaurs.

Class:	Cephalopoda
Order:	Phylloceratida
Habitat:	Marine
Distribution:	Global
Time scale:	Jurassic–Cretaceous

123

Harpoceras sp.

Ammonites were an incredibly successful and diverse group of ancient molluscs. And although we only have one coil-shelled extant cephalopod, the nautilus, to use as an analogue, we can still make some guesses as to the behaviour of the extinct cephalopods with coiled external shells. *Harpoceras* was a very laterally compressed Jurassic ammonite. This feature appears to indicate swiftness, and, thus, likely a predatory habit for *Harpoceras*. The specimen seen here has been painted to contrast two other details of the shell: the fractal-like pattern on the left shows the suture marking the internal septa that separate the chambers and the sinuous radiating curves at bottom centre represent the external ornament of the shell. Both of these patterns, especially the sutures, are useful in determining ammonite species.

Class:	Cephalopoda
Order:	Ammonitida
Habitat:	Marine
Distribution:	The Americas, Europe, Asia
Time scale:	Jurassic

Parkinsonia sp.

The geometric precision of spiral shelled molluscs like snails and ammonites has fascinated people for ages. Very simple mathematical rules are employed by the animals to produce deceptively complex patterns. Slight variations to the basic maths behind these designs results in seemingly boundless variation. And since these slight differences are governed by the DNA of each species, attention to these patterns can define and separate different species. This specimen has been polished in order to remove the outer detail and expose the intricate, species-specific sutures useful in ammonite identification. The original external ornamentation can be seen as ribbing on the unpolished centre of the coil called the umbilicus. The extent of polish suggests that the polishing was done artistically rather than purely for research.

Class:	Cephalopoda
Order:	Ammonitida
Habitat:	Marine
Distribution:	Canada & Europe
Time scale:	Jurassic

Amaltheus sp.

The spiral shell of the *Amaltheus* is highly compressed. Sickle-shaped ribs ornament its flanks, though some fossils bear fine twisting ribs. *Amaltheus* had a narrow keel that resembled a rope: the keel indicates that unlike *Gastrioceras*, *Amaltheus* was a faster swimmer. *Amaltheus* fossils, which have been found all over the world, may possibly have evolved from *Phylloceras*, a rather more streamlined ammonite that could swim at impressive speeds by using jet propulsion. Ammonites like *Amaltheus* became extinct at much the same time as the non-avian dinosaurs, at the end of the Mesozoic era around 65 million years ago. As with the non-avian dinosaurs, the reasons for their extinction remain subject to several theories.

Class:	Cephalopoda
Order:	Ammonitida
Habitat:	Marine
Distribution:	Canada & Europe
Time scale:	Jurassic

Hoplites sp.

Hoplites had slightly involute (tightly coiled) shell shaped like a disc with an open umbilicus (the area around which the shell coils) at its centre. Whorls that were oval or square in cross-section characterized the shell, with ribs that branched out towards – but did not continue along – the venter (outer edge). The ribs themselves were thick and close together, spreading out from the tubercles lying on the edge of the umbilicus. *Hoplites* lived at the end of the Early Cretaceous and were found mainly in Europe. Their typical shell size was around 8cm (3¼in). Their name, which reflects their rather aggressive appearance, recalls the heavily armed hoplite warriors of Ancient Greece.

Class:	Cephalopoda
Order:	Ammonitida
Habitat:	Marine
Distribution:	Europe
Time scale:	Early Cretaceous

Psiloceras sp.

Psiloceras, an ammonite that has been found all over the world, often acts as an indication of the Early Jurassic Period, when dinosaurs were abundant on earth. *Psiloceras* was small, only about 7cm (2¾in) in diameter, and may have evolved from the family Phylloceratidae. *Psiloceras* retained the same type of suture where the chambers connected. One particular site where *Psiloceras* has been found in north Somerset, in southwest England, have a deep pink colour and they are preserved by being crushed flat in shale. Although the grey shale in which they are found has flattened them, it was capable of preserving the gorgeous colours of their mother-of-pearl of nacre.

Class:	Cephalopoda
Order:	Ammonitida
Habitat:	Marine
Distribution:	Europe, the Americas, New Zealand
Time scale:	Early Jurassic

Ammonite

Ammonites are common fossils and are easily recognized. Extinct relatives of todays octopuses and squids, they lived in the seas in huge numbers right up until the end of the Cretaceous Period. Some of them were tiny while others grew to a huge size, with shells over 1m (3¼ft) across. Ammonite shells show their relationship to other, more simple molluscs such as sea snails. Unlike these creatures, however, ammonites swam in open water. Their shells had interconnected chambers, which they used to control their buoyancy and hold their position in the water. Fossils such as this one, where the shell has been cut in two and polished, show those chambers. The animal's body filled the largest, open chamber at the end of the spiral, filled here by the fossil of another, smaller ammonite.

Class:	Cephalopoda
Order:	Ammonitida
Habitat:	Marine
Distribution:	Global
Time scale:	Silurian–Cretaceous

Pinacoceras metternichi

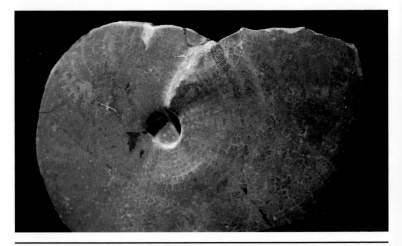

Pinacoceras was the largest known genus among the ammonoids that lived in the Triassic period. For example, *Pinacoceras metternichi*, pictured here, was a notable giant – its shell could grow up to 1.5m (5ft) tall. Despite its size, though, the shell was very slim and flat, consisting of narrow, spirally curled whorls. The whorls were sharp-edged on the outer edge. *Pinacoceras*' narrow mouth aperture was sited quite high up on the shell and the suture (the line where the chambers connected) was distinctly ammonitic and featured a complex design. The outside surface of the shell was very smooth. *Pinacoceras* lived in many parts of Europe and Asia and more than 20 species have been found.

Class:	Cephalopoda
Order:	Ceratitida
Habitat:	Marine
Distribution:	Global
Time scale:	Late Triassic

Hamites sp.

Ammonites were an incredibly diverse group of marine molluscs that enjoyed global distribution and existed for hundreds of millions of years before dying out 65 million years ago. The majority of them had spiral shells reminiscent of today's *Nautilus*, but ammonites were actually more closely related to the octopuses. Many separate lines of ammonites developed shells with odd spirals or other bizarre twists that somehow related to their particular mode of life. This *Hamites*, shown in the split concretion in which it was found, shows the peculiar 'hairpin' bend of its kind. Since modern shelled cephalopods have shapes unlike *Hamites,* it is difficult to say what the advantage was in this shell type. Along the edge of the rocks one can see another small ammonite of the more common morphology.

Class:	Cephalopoda
Order:	Ammonitida
Habitat:	Marine
Distribution:	Global
Time scale:	Cretaceous

Octopus sp.

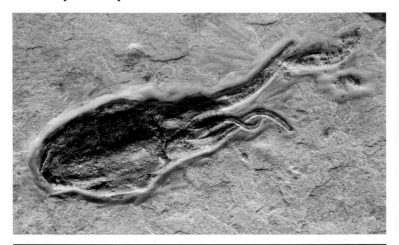

Octopus fossils are rare. The reason for this is twofold. Firstly, octopuses themselves have probably never been common – unlike their relatives the squid, they are solitary hunters that live and feed alone rather than in shoals. Secondly, octopuses have soft bodies with almost no hard structures. The only truly tough part of an octopus is its parrot-like beak, which it uses for biting and killing prey. Soft-bodied animals rarely fossilize well. Octopuses have lived on Earth for well over 100 million years, and they had very similar-looking ancestors that were already widespread by the start of the Jurassic Period, 208 million years ago. This Jurassic specimen shows extraordinary preservation, not only in that its soft anatomy is represented, but that it is in three dimensions.

Class:	Cephalopoda
Order:	Octopoda
Habitat:	Marine
Distribution:	Global
Time scale:	Cretaceous–Recent

Belemnite

Belemnites were close relatives of modern-day squid and cuttlefish. Like them, they had 10 flexible tentacles for capturing prey and a sac in their bodies that contained ink, which they squirted out to create a 'smoke screen' to confuse predators. Belemnites had soft bodies built around a hard bullet-shaped core. It was this that was usually preserved as a fossil – the rest of the animal usually rotted away. Belemnites were very common in Jurassic and Cretaceous seas, as were their relatives the ammonites. Both groups died out at the end of the Cretaceous Period, killed off in the mass extinction event that also wiped out the non-avian dinosaurs.

Class:	Cephalopoda
Order:	Belemnitida
Habitat:	Marine
Distribution:	Global
Time scale:	Early Carboniferous–Cretaceous

Acanthoteuthis sp. (arm hooks)

*A*cantho*teuthis* was a primitive squid relative called a belemnite. Like modern squid, it had tentacles, but they were lined with hooks instead of suckers. These hooks give *Acanthoteuthis* its name, which means ' thorny squid'. Fossils of *Acanthoteuthis* are rare, not because the animal itself was rare but because it had a soft body that was normally eaten by other animals soon after it died. Even when that did not happen, it usually rotted away before it could be covered by sediment to form a fossil. Although the few fossils of this squid that are known usually show single animals, it is likely that *Acanthoteuthis* lived and hunted in shoals, as most squid do today. *Acanthoteuthis* grew to about 50cm (20in) long.

Class:	Cephalopoda
Order:	Belemnitida
Habitat:	Marine
Distribution:	Europe
Time scale:	Jurassic

Squid (pen)

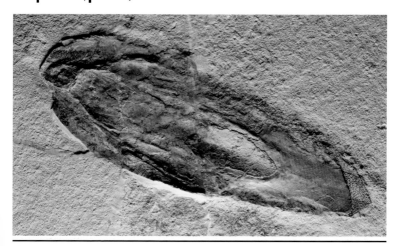

Squids, or teuthids, are found in the fossil record at least as far back as the Jurassic. Their fossil record is very poor, but this is probably due more to the fact that they are hard to preserve than they themselves were uncommon. The bodies of squids are very soft and decay quickly, as do the only parts of a squid that one might call 'hard': the beak and the cuttlebone (pen). This internal stiffening element is what is preserved in this Jurassic specimen from Germany. Despite this dearth of preservation potential, there are extraordinary deposits that had just the right conditions for squid preservation, and even the most delicate parts of squids have been preserved: tentacles, arm hooks, eyes, mantle muscles, intestines, blood vessels, gills and even ink sacs with the ink preserved.

Class:	Cephalopoda
Order:	Teuthida
Habitat:	Marine
Distribution:	Global
Time scale:	Jurassic Period–Recent

Marrella splendens

Finding complete soft-bodied animal specimens is rare in palaeontology, making this fragile arthropod a part of one of the most important fossil collections in the world. Since its discovery in 1909 by Charles Walcott (1850–1927), over 25,000 examples of *Marrella splendens* have been found in Canada's Burgess Shale. Measuring 2.4mm (⅒in) to 2.5cm (1in) long, these tiny 'lace crabs', as they are sometimes called, have a segmented triangular body with two pairs of antennae and wispy gills. In 2004, an example was even found in the process of moulting (shedding its tough outer exoskeleton in order to grow). It had long been believed that ancient arthropods moulted in just the same way as their modern-day descendants. However, this was the first time it had been seen in the fossil record.

Order:	Mimetasterida
Family:	Marrellidae
Habitat:	Marine
Distribution:	Canada's Burgess Shale
Time scale:	Middle Cambrian

Paradoxides sp.

Paradoxides was the largest known trilobite of the Cambrian Era, measuring nearly 1m (3¼ft) in length. The average, as opposed to the giant, size of Paradoxides was about 20cm (8in) in length. *Paradoxides* had a large head and long thorax (middle section). As *Paradoxides* grew, fresh sections were added to the thorax, which tapered gradually to the small tail. The total number of sections was between 19 and 21. The central part of the head (glabella) was pear-shaped and had a large inflated front lobe, and the cheeks extended to the long spines. *Paradoxides* lived mainly in Europe, North Africa and the United States, but in the Middle Cambrian it was widespread across the world.

Class:	Trilobita
Order:	Redlichiida
Habitat:	Marine
Distribution:	Northern Hemisphere
Time scale:	Middle Cambrian

Greenops sp.

*G*reenops, which inhabited the shallows of warm seas across the world, was a small oval-shaped trilobite typically measuring around 4.5cm (1¾in) in length. Its head was convex and extended forwards. The eyes were large, protruding and arch-shaped. The small tail, which had deep furrows, was smaller than the head and tucked under the head when *Greenops* rolled itself up. *Greenops'* ability to roll up was probably a defence measure against predators. Many *Greenops* fossils have been found in a tight rolled-up state. The spines on its pygidium, or tail, projected beyond the margin of the cephalon, or head, when the animal was enrolled.

Class:	Trilobita
Order:	Phacopida
Habitat:	Marine
Distribution:	Europe, Africa and The Americas
Time scale:	Devonian

Calymene sp.

Like many other trilobites – a name derived from the three vertical lobes that comprised their bodies – *Calymene* had a strong convex exoskeleton. *Calymene's* eyes were rather small, and lay close to the heavily channelled glabella at the centre of its head. *Calymene* was usually about 7cm (2¾in) long, and their fossils have been found all over the world. Quite probably, *Calymene* was not a good swimmer and some palaeontologists believe that this trilobite walked on the sea bed. *Calymene* was another trilobite that avoided danger by rolling itself into a ball.

Class:	Trilobita
Order:	Phacopida
Habitat:	Marine
Distribution:	Global
Time scale:	Ordovician–Devonian

Asaphus expansus

*A*saphus was a large trilobite species, growing to well over 60cm (23½in) long. The group to which *Asaphus* belongs gave rise to more than 30 new species of trilobite in just two million years – a very short time in terms of evolution. This evolutionary explosion was driven by changes in the habitat in which these trilobites lived. Among them was the famous *Asaphus kowalewski*, which had its eyes up on long stalks. These enabled it to spend most of its time hidden beneath the mud but still be able to look out for predators. This specimen of *Asaphus expansus* is probably a shed moult as evidenced by the gap between the head and the rest of the body – this is where the animal vacated its old exoskeleton.

Order:	Asaphida
Family:	Asaphidae
Habitat:	Marine
Distribution:	Europe, South America and Asia
Time scale:	Ordovician

Phacops rana

P*hacops rana* was a trilobite that lived in shallow seas of the Devonian. It is notable for its large eyes, which were mounted on raised turrets, giving it excellent all-round vision. The word 'rana' is Latin for frog and was given to this species because of its bulging eyes and semicircular head, which is reminiscent of the face of a frog. *Phacops rana* was quite large for a trilobite, growing up to about 15cm (6in) long. It is a very common fossil in parts of North America and was made the official state fossil of Pennsylvania in 1988. In life, *Phacops rana* was able to roll up into a ball for protection from predators and many fossils of it, like the one pictured here, show it in this position.

Order:	Phacopida
Family:	Phacopidae
Habitat:	Marine
Distribution:	Global
Time scale:	Devonian

Dalmanites sp.

*D*almanites trilobites have a distinctive body shape – rounded at the front and tapering to a point at the back. They were widespread on the sea floor around 480 million years ago, rummaging through the sediment for worms and other small prey. *Dalmanites* had well-developed compound eyes, which can still be seen clearly on some fossil examples. Often, these trilobites are found as incomplete specimens. Their bodies quickly broke apart after death, but disarticulate specimens may also represent moults. Only those that were rapidly buried by sediment could survive intact. Most *Dalmanites* species were relatively small, rarely growing to more than 5cm (2in) long.

Order:	Phacopida
Family:	Dalmanitidae
Habitat:	Marine
Distribution:	Global
Time scale:	Ordivician–Devonian

Dalmanitina socialis

*D*almanitina belonged to a group of trilobites, the Phacopida, which is of
considerable stratigraphic importance. Their fossils have helped date the strata
or layers of ancient rock in which they have been found. The unusual structure of
the phacopid exoskeleton makes them easy to identify, even when only fragments
remain. *Dalmanitina* had the characteristic slightly convex exoskeleton. The head
was curved or semicircular and the glabella (central section) was pear-shaped.
Dalmanitina's body was divided into eleven segments. The tail was quite large and
had six pairs of ribs on the side lobes. *Dalmanitina* was widespread throughout the
world until it died on a vast scale in an extinction possibly caused by an ice age
between 440 and 450 million years ago.

Class:	Trilobita
Order:	Phacopida
Habitat:	Marine
Distribution:	Global
Time scale:	Ordovocian

Ellipsocephalus hoffi

Ellipsocephalus hoffi was one of the first trilobites described by the German palaeontologist Friedrich von Schlotheim (1764–1832) in 1823. It became one of the most important trilobites of the Cambrian Era to be discovered in the Jince region of the present-day Czech Republic. Unlike too many trilobite species, the Jince discoveries were well preserved, and entire groups have been found whole and undamaged. *Ellipsocephalus* usually measured around 3cm (1¼in) and was oval-shaped with a semicircular head. The glabella at the centre of the head was smooth, broadening at the front. Apart from in Europe, fossils of *Ellipsocephalus* have been found in North America. The marks at the rear of the specimen shown here look like legs but are actually tool marks.

Class:	Trilobita
Order:	Ptychopariida
Habitat:	Marine
Distribution:	Europe and North America
Time scale:	Cambrian

Asaphiscus wheeleri

A *saphiscus wheeleri* is a common trilobite fossil found in the Wheeler Shale in Millard County, Utah. The trilobite fauna of the Wheeler is world-renowned and some species, like *Elrathia kingi*, are so common they have become inexpensive commercial fossils. The trilobites were arthropods, and just like the others in this group, they moulted their hard exoskeleton periodically to accommodate their growth. This specimen of *Asaphiscus* is a shed exoskeleton, evidenced by the displaced central part of the cephalon ('head'). The sides of the cephalon ('free cheeks') are missing which, along with the separation of the cephalon from the rest of the body, marks two other sutures (seams along which the exoskeleton split in the first stages of moulting, allowing the trilobite to free itself from its old covering).

Order:	Ptychopariida
Family:	Asaphiscidae
Habitat:	Marine
Distribution:	Utah, USA
Time scale:	Middle Cambrian

145

Eophrynus sp.

*E*ophrynus belongs to the Class Arachnida (spiders and their relatives). Like other members of the Order Trigonatarbida, *Eophrynus* did not posses the silk glands with which spiders spin their webs and trap their prey, nor the poison glands employed to kill prey. Unlike its relative, the scorpion, *Eophrynus* did not carry rigid armour for protection and, except for the abdomen, its body was unsegmented. *Eophrynus* had a long oval shape overall and possessed a triangular cephalothorax (a combined head and trunk). Its nine abdominal segments could be seen from its upper surface. The surface of *Eophrynus* was covered with a vast number of tubercles (small rounded projections).

Class:	Arachnida
Order:	Trigonotarbida
Habitat:	Terrestrial
Distribution:	England
Time scale:	Upper Carboniferous

Eurypterus sp.

*E*urypterus was a sea scorpion. It was a strong swimmer, with a very long body that narrowed at the back down to slender point at the telson (the often pointed tail of some arthropods, like in horseshoe crabs). The joined head and thorax was nearly square-shaped, with its corners rounded. At each side of the head, *Eurypterus* had two pairs of eyes: one compound the other – in the middle of the first pair – were simple eyes. *Eurypterus'* front walking legs were covered in spines. It swimming legs were well adapted for their task, and were wider and paddle-shaped. *Eurypterus* lived over a wide area – in Europe, North America and Asia.

Class:	Merostomata
Order:	Eurypterida
Habitat:	Aquatic
Distribution:	Northern Hemisphere
Time scale:	Silurian–Devonian

Euproops sp.

Euproops was a very early member of the horseshoe crab group. It had a carapace shaped like a shield over its front. There were seven fused segments making up the abdomen, which featured a fringe of spines around the edges. The last segment was different, though – it was a long telson (tail spine) used mainly to right the animal when it was flipped over. *Euproops* is thought to have lived on land. Its average length measured 4cm (1½in). The sea scorpion has been found in North America and Europe. Chelicerates such as *Euproops* were arthropods but, unlike other arthropods, possessed no antennae. Some scientists think that this tiny horseshoe crab was terrestrial and hid among the leaves of certain early land plants.

Class:	Merostomata
Order:	Xiphosurida
Habitat:	Aquatic
Distribution:	Europe and North America
Time scale:	Carboniferous

Mesolimulus sp.

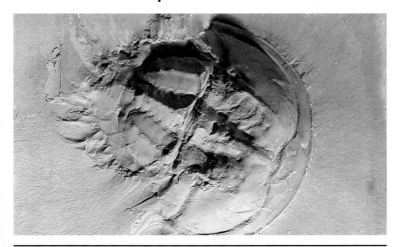

*M*esolimulus, whose fossils have been discovered in Europe and the Middle East, was an early horseshoe crab. Like its modern relatives, *Mesolimulus* had a horse-shoe shaped carapace, and on this head shield were its small, widely spaced eyes. *Mesolimulus*, which measured an average 12cm (4¾in) long, had three vertical ridges along its carapace and a fused abdomen. Although the abdomen had no segments, there were six short spines around its rim, and it was connected to the carapace and the long, sharp tail spine. The similarity between this ancient species and the few surviving horseshoe crab species of today is striking, giving *Mesolimulus* common recognition as a 'living fossil'.

Class:	Merostomata
Order:	Xiphosurida
Habitat:	Lagoons
Distribution:	Europe
Time scale:	Jurassic

Insect (in amber)

Although amber is often cut, polished and set in jewellery, it is not a gem stone. In fact, it is a fossilized reminder of the forests of our ancient world. Amber, itself a fossil, was formed from the resin of trees. Deposits of amber have been known and admired since the time of the Neolithic people, who used it to make jewellery. The most famous amber deposits are found in the Baltic area, and it is these examples that typically contain insects, leaves or flowers – trapped by the tree's sticky resin, millions of years ago. In the film *Jurassic Park* the park's creator John Hammond famously used the blood of a mosquito that had fed on a dinosaur and then been trapped in amber to create the inhabitants of his dinosaur theme park, but this is pure fantasy.

Order:	Many
Family:	Many
Habitat:	Various
Distribution:	Global
Time scale:	From the Early Cretaceous but very possibly older

Insect and Tetrapod (tracks)

Most obvious on this small slab of rock from the Permian of Abilene, Texas, are the impressions of small early tetrapod tracks near the bottom center. But subtle zipper-like marks on the right side are the delicate trackways of insects or other small arthropods. Even fainter marks show the path of some other small invertebrate. The red and green coloration may just be staining and has nothing to do with the tracks themselves, but probably relates to the mud that formed the rock and how it lithified or weathered. The tracks and trails preserved on this rock are extending beyond the surface, as opposed to being impressed into it, because this rock comes from the sediments that covered and filled the actual track-marked surface trod by these animals.

Order:	n/a
Family :	n/a
Habitat:	n/a
Distribution:	Global
Time scale:	Devonian–Recent (both insects and tetrapods)

Insect

Tar pits, or more accurately asphalt pits, conjure images of life and death struggles between huge mammoths and ravenous sabre-toothed cats. But asphalt does not discriminate, and if asphalt fossil hunters also resist discriminating towards the big and bony, they notice a host of smaller victims too. Like sticky tree resin, asphalt can immobilize and engulf insects and other small invertebrates. Like many of the other larger victims, this Pleistocene beetle may have approached the asphalt thinking it was water and got trapped after touching it. But rotting carcasses attract scavengers of all kinds, with insects making up a large portion of them, and this one may have been attracted to a free meal only to pay with its life. A gust of wind or other causes also cannot be ruled out.

Order:	Coleoptera
Family:	n/a
Habitat:	Various terrestrial & aquatic
Distribution:	Global
Time scale:	Early Permian–Recent

Chresmoda sp.

This little creature was one of the earliest pond skaters (insects that skim across the surfaces of ponds and other areas of still fresh water). *Chresmoda*, like modern pond skaters, could fly. This enabled it to move on and find new places to live if the area of water it had been living on dried up. Although it looks delicate, *Chresmoda* lived by killing and eating other flying insects that crashed into the areas of water it inhabited. It picked up the vibrations and ripples they made as they struggled with its legs, quickly skating over to stab its prey with its pointed proboscis and suck it dry.

Class:	Insecta
Family:	Chresmodidae
Habitat:	Fresh water
Distribution:	Europe
Time scale:	Jurassic

Eocicada sp.

This fossil shows an insect that would not look out of place in a forest today. *Eocicada*, shown here, could fly like other cicadas, but probably spent most of its time lying still amongst leaves. It was entirely herbivorous, living on plant sap. Modern cicadas are known for their extremely loud 'songs', composed of long buzzes and chirruping sounds. It seems almost certain that *Eocicada* would have made similar songs to attract mates, and it is not hard to imagine its loud buzzing echoing through the forests when non-avian dinosaurs still roamed the earth.

Order:	Homoptera
Family:	Cicadidae
Habitat:	Forests
Distribution:	Europe
Time scale:	Jurassic–Cretaceous

Isophlebia sp.

Isophlebia was a damselfly that lived at the same time as the heyday of the gigantic long-necked sauropod dinosaurs. Like its modern descendants, it was an aerial hunter, grabbing smaller flying insects on the wing. *Isophlebia* spent most of its life as a wingless larva living underwater. It lived as an adult for only a few months, long enough to find a mate and lay eggs before dying. *Isophlebia* was closely related to the dragonflies that also existed at this time. Unlike them, however, it could fold its wings up over its back – a feature that still separates damselflies from dragonflies today. *Isophlebia* laid its eggs on the leaves of water plants and was rarely found far from fresh water.

Class:	Insecta
Order:	Odonata
Habitat:	Near fresh water
Distribution:	Europe
Time scale:	Jurassic

Insects

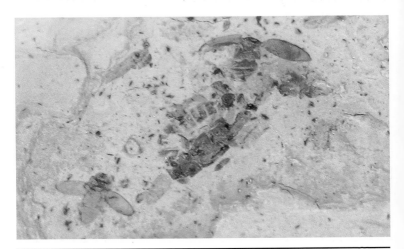

Tiny insect fossils are famously and exquisitely preserved in amber. For excellent preservation of miniscule and fragile insects to occur in rock, burial must be quick, but gentle, and the sedimentary particles must be tiny. Such is the case for some areas of the Eocene Green River Formation in Colorado, where these insects were found. Two beetles are obvious on this slab – both with their elytra, or wing cases, open as if flying. Also preserved is a type of fly on the left and other unidentified insect parts. The tiny black bits are plant debris. Insect fossils are found here by the thousands, but why they died in such numbers is unknown. They might have all been killed at once or their dead bodies may have accumulated over time.

Order:	Coleoptera (beetles); Diptera (robber fly); n/a(others)
Family :	n/a (beetles); Asilidae (robber fly), n/a (others)
Habitat:	Near lakes (these specimens)
Distribution:	Global (beetles and robber flies)
Time scale:	Eocene (these specimens)

Ephemeropsis trisetalis (nymph)

Mayflies were one of the first insect groups to take flight, with a history probably extending all the way back to the Carboniferous, and are the most basal group of living flying insects. The naiads, or nymphs, are aquatic and breath through gills on their sides. This naiad comes from the famous Liaoning deposits of China, renowned for their incredible preservation, diversity and abundance of fossils, including feathered dinosaurs. This specimen shows the three caudal filaments from which its species name derives, that are actually more gills. Its genus name comes from the fact that mayflies live a very brief life, where the only function of adults is to disperse and reproduce. After the nymph eats, grows and metamorphoses, the adult form, which never eats, tries to find a partner and mate before dying.

Order:	Ephemeroptera
Family :	Hexagenitidae
Habitat:	Fresh water
Distribution:	Liaoning, China
Time scale:	Early Cretaceous

Libellula doris (larva)

*L*ibellula doris is the name given to a type of fossilized dragonfly larva. Like the larvae of dragonflies today, it lived in fresh water, hunting other insects, tadpoles and small fish. *Libellula doris* caught its food by lying in wait, motionless, then striking suddenly when prey ventured too near. It grabbed them in its formidable mouth parts, which it could shoot out suddenly in front of its head. Normally, dragonfly larvae live underwater for a couple of years until they are ready to change into adults. To do this, they crawl up the stem of a plant until they are out of the water. They then break out of their old skin and pump up their bodies and wings with blood before flying away.

Class:	Insecta
Order :	Odonata
Habitat:	Near freshwater
Distribution:	Europe
Time scale:	Miocene

Cyclorraphan (maggot)

Cyclorrhaphans are a group of flies comprising around 65,000 species. This fly larva, or maggot, comes from Douglass Pass, Colorado, a prolific producer of fossil insects: around 90% of the fossils can be insects, with plants, fish and feathers making up the remaining 10%. And although no mammal fossils have yet been unearthed at this particular site, it is clear they lived in the area at the time. Backwards-pointing spines suggest that this larva is a botfly: a notorious cyclorraphan parasite of large grazing mammals. Such spines enable bot larvae to anchor themselves in a part of their host – nostrils, guts, or under its skin. Alternatively, these larvae could have been carrion feeders. Some layers at this site contain only these maggots, but this anomalous accumulation remains unexplained.

Order:	Diptera
Family :	Oestridae, Calliphoridae, or Sarcophagidae
Habitat:	Terrestrial
Distribution:	Global
Time scale:	Cretaceous–Recent (this specimen is Eocene)

Termite (coprolites)

Plants and insects have interacted closely for nearly their entire histories. These interactions cover the whole spectrum, going from mutually beneficial symbioses to commensalism, where insects live with but don not injure their host, to parasitism and herbivory that harm or even kill the plant. Some of the earliest anatomically preserved plant fossils from the Devonian of Scotland clearly show damage due to the sucking mouthparts of insects. In this Tertiary piece of petrified incense cedar wood (*Libocedrus* sp.) from Buellton, California, the galleries and frass of termites can be seen. Under magnification, the hexagonal cross-sections of these tiny coprolites indicate their origin. Insects and other arthropods have been tunneling into wood at least since the Carboniferous, shortly after wood first shows up in the fossil record.

Order:	Isoptera
Family:	n/a
Habitat:	In wood
Distribution:	Global
Time scale:	Early Cretaceous–Recent

Aeger insignis

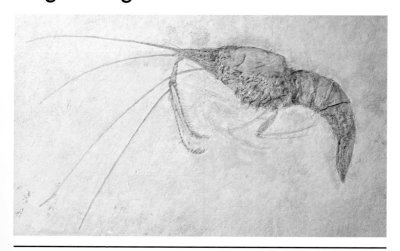

The lithographic limestones of Germany have attracted printers and palaeontologists for a long time. Both have sought the properties afforded by the fine grain
sediments the limestone is made from: printers use it to capture the fine detail of
drawings, while palaeontologists are interested in the fine details of the fossils that it
preserves. The calm lagoonal setting of the limestone's deposition allowed for slow,
gentle burials that conserved the delicate features of such anatomical fragility as
feathers, dragonfly wings and shrimp antennae. *Aeger* is the most abundant decapod
from the lithographic limestones and preserves delicate legs and feelers in the exact
position of the living animal. This amazing preservation was aided by the fact that the
depths of the lagoon were anoxic and thus decay-causing organisms were absent.

Class:	Malacostraca
Order:	Decapoda
Habitat:	Lagoons
Distribution:	Germany
Time scale:	Late Jurassic

Pagurus sp.

This fossil shows a creature many people would be familiar with today. *Pagurus* was a hermit crab, a small crustacean that protected its body by hiding it inside the empty shell of a dead sea snail. *Pagurus* lived like most crabs, scavenging on the sea floor and around the tide line on beaches for dead fish and other bits of organic matter. They creep around when they feel safe, and pull themselves back inside its shell as soon as they feel threatened and block up the entrance with its pincers and legs. Most crustaceans regularly shed their exoskeletons as they grow. *Pagurus* did this but also changed the shell it carried, moving into a larger one every time it moulted.

Order:	Decapoda
Family:	Paguridae
Habitat:	Coastal marine
Distribution:	Global
Time scale:	Cretaceous–Recent

Palaeocarpilius sp.

During the Eocene Period, many crabs lived on Earth and, even that many millions of years ago, they were starting to resemble the crabs we know today. One of them, *Palaeocarpilius*, was a crab we could easily recognize. Measuring 6cm (2¼in) in length, the carapace on its back was oval-shaped with an extended front and spiny edges along the front and sides. *Palaeocarpilius* had six long, strong legs and, of its two powerful claws, the right was larger than the left. *Palaeocarpilius'* all-embracing shell curved down and backwards along its underside and ended beneath the narrow plate in front of its mouth. This crab inhabited the shorelines of Africa and Europe, preferring the more tropical areas of these continents.

Class:	Malacostraca
Order:	Decapoda
Habitat:	Marine
Distribution:	Africa and Europe
Time scale:	Eocene–Miocene

Phalangites priscus (larva)

W hen this fossil was first discovered and described, it was thought to be an entirely new type of crustacean. Since then, however, scientists have changed their minds. It is now thought to be the larval form of a fossil decapod – exactly which one we will probably never know. Decapods are crustaceans with ten limbs. Modern decapods include lobsters and crabs. Crustaceans go through several stages in their lives, and as larvae (young animals) they often look very different from their parents. Although *Phalangites priscus* had long spidery legs, that does not necessarily mean that the animal it grew into had long legs as well. What scientists can tell from this fossil is that *Phalangites priscus* probably lived in deep waters as it has long antennae – useful for travelling over the seabed at depths where it is too dark to see.

Class:	Malacostraca
Order:	Decapoda
Habitat:	Deep seabed
Distribution:	Europe
Time scale:	Jurassic

Maja orbignyi

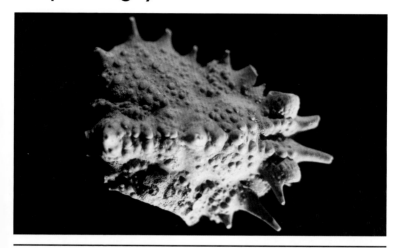

The family Majidae contains many hundreds of modern species (including the commercially exploited Canadian snow crab) but are more commonly referred to as spider crabs in reference to their long slender legs. The group is widely distributed in marine waters all around the world today and may have its origins back in the Cretaceous, although the oldest known true majids are Eocene. The spines and other ornamentation on the carapace of this fossil specimen of *Maja orbignyi* is typical for the group and help to disguise these slow-moving crabs amongst the detritus and other floating bits in its seabed home. Decorator crabs, which are also in this family, take this camouflage to an extreme level by adorning themselves with organic debris of all kinds, rendering them practically invisible.

Order:	Decapoda
Family:	Majidae
Habitat:	Marine
Distribution:	Global
Time scale:	Eocene Period–Recent (Majidae)

Crab

Crabs have existed on Earth for many millions of years. To the majority of people they are the most familiar and easily recognized of all crustaceans, and this is true for their fossils as well. Many fossils of crabs show the whole animal, complete with legs and pincers, but this one shows just the carapace. Perhaps surprisingly, fossils like this (of empty carapaces) are not uncommon. This is because crabs, like all crustaceans, regularly shed their shells as they grow. Fossils with legs and claws were usually formed from animals that died, but fossils like this one were created when their empty shells were buried. Like most fossils, the substances that made up the shell were replaced over millions of years by minerals from the rock that formed around it.

Order:	Decapoda
Infraorder:	Brachyura
Habitat:	Seabed
Distribution:	Global
Time scale:	Jurassic–Recent

Acanthochirana cenomanica

The shrimp *Acanthochirana* was confined to a single country, Lebanon, in the eastern Mediterranean. Most members of the Penacidae family, to which *Acanthochirana* belonged, were fully fledged marine creatures and lived far out to sea. Some, though, preferred the fresh or brackish water to be found nearer the Mediterranean coast. The typical length of this shrimp was around 3.5cm (1⅜in) and it had a short cylindrical carapace. The small size was deceptive. *Acanthochirana* could appear much larger because of its very long antennae, which were longer than its body. The third pair of antennae was the longest, the first pair being much shorter and featuring flexible joints. *Acanthochirana's* tail spine was shaped like a spindle and the tail ended in fan-shaped segments.

Class:	Malacostraca
Order:	Decapoda
Habitat:	Marine
Distribution:	Lebanon
Time scale:	Jurassic

Ceratiocaris sp.

This little creature swam in the seas over what is now Europe 430 million years ago. It belonged to a now long-extinct group of crustaceans known as the pod shrimps. *Ceratiocaris* had a soft body, the front half of which was protected by a hard carapace (shell). Most fossils of this animal show only the carapace but, occasionally, the soft parts are preserved as well. *Ceratiocaris* was one of the earliest animals to develop a hard shell, and it is commonly found in rocks laid down during the Silurian Period.

Order:	Archaeostraca
Family:	Ceratiocaridae
Habitat:	Marine
Distribution:	Europe
Time scale:	Ordovician–Devonian

Carpopenaeus callirostris

Unlike many fossils – which show long-lost forms of life or are too unclear to identify easily – there is no mistaking what this creature was. *Carpopenaeus callirostris* was a shrimp that lived in deep open water in the oceans at around the same time as *Tyrannosaurus rex* roamed the earth. Although it lived millions of years ago, *Carpopenaeus callirostris* looked very much like many deep-water shrimps around today. Like them, it had extremely long antennae for detecting other creatures in the darkness. It also had little paddles, known as swimmerets, beneath the rear half of its body to propel it through the water. This shrimp's front limbs were long and covered with little spikes. It probably used these to grab the smaller creatures on which it fed.

Class:	Malacostraca
Order:	Decapoda
Habitat:	Deep sea
Distribution:	Global
Time scale:	Late Cretaceous

Palaeosculda sp.

*P*alaeosculda was a type of mantis shrimp that lived near the end of the Cretaceous Period, when the last of the great non-avian dinosaurs were roaming the land. Like the mantis shrimps of today, it lived on the sea bottom, spending much of its time sheltering in a burrow that it dug in the sand. *Palaeosculda* was a relative of today's crabs and lobsters, but unlike them it had clubs on the ends of its two front limbs rather than pincers. These were used to literally punch prey to death. Mantis shrimps today have such powerful punches they can easily break a man's finger. *Palaeosculda* grew to just over 10cm (4in) long and preferred the warm shallows of tropical seas.

Class:	Malacostraca
Order:	Stomatopoda
Habitat:	Sandy sea floor
Distribution:	Global
Time scale:	Late Cretaceous

Cycleryon sp.

Cycleryon was a marine crustacean that roamed the seabed around 150 million years ago. It was related to the polychelid lobsters, a little-known and poorly understood group of crustaceans that still inhabit the deep sea today. Like modern polychelids, *Cycleryon* had pincers on the ends of all 10 of its limbs, not just on the first two like most other lobsters. Its pincers were delicate, suggesting that it fed on small or soft food items. It was probably a scavenger, picking away at the bodies of dead fish and other animals that drifted down from the water above. *Cycleryon* had very small eyes and was probably almost blind. In the deep sea where it lived, vision would have been useless anyway, as everything would have been shrouded in darkness, far from the rays of the sun.

Class:	Malacostraca
Order:	Decapoda
Habitat:	Deep sea
Distribution:	Europe
Time scale:	Jurassic

Eryon arctiformis

*E*ryon *arctiformis* is often referred to as a lobster, but its body form is practically intermediate between lobsters and crabs. Fossils of *Eryon* and its close relatives have been found in Late Jurassic lithographic limestones of Germany. They appear to be related to polychelid crustaceans, deep sea crustaceans that live at depths between 100m (328ft) and 5000m (16,400ft). Because of the great depth of their preferred habitat, and despite their apparently being widespread, the whole family went unnoticed until the famous Challenger expedition of the late nineteenth century dredged some up. Their similarity to eryonids was immediately noted and although they are certainly distinct, groups may be closely related and clarify not only the classification of eryonids but might illuminate their behaviour as well.

Class:	Malacostraca
Order:	Decapoda
Habitat:	Marine
Distribution:	Europe
Time scale:	Jurassic

Homarus hakelensis sp.

This lobster swam in the shallow Tethys Sea, which covered much of Europe during the Cretaceous Period. Although it lived more than 100 million years ago, it looked similar to many lobsters that are alive today. *Homarus hakelensis* was quite a small animal, growing to a little over 10cm (4in) long. It used its hefty pincers both for defence and to tear chunks off the dead fish and other animals it scavenged for on the seabed. *Homarus hakelensis* could see, but found its way around mainly by touch, using the long antennae on its head to feel the seabed in front of it. If it needed to escape predators, it could move quickly through the water, shooting backwards suddenly with a powerful flick of its tail.

Class:	Malacostraca
Order:	Decapoda
Habitat:	Seabed
Distribution:	Europe
Time scale:	Cretaceous

Linuparus sp.

The spiny lobsters (Family Palinuridae) are a common sight in tropical waters and have their earliest recognized origins back in the Cretaceous. They are distinguished from true lobsters (Family Nephropidae) by their heavy spiny antennae and complete lack of claws. They are not closely related to the true lobsters and arrived at a similar body form independently. The etymology of the name *Linuparus* is extremely uncommon in biological nomenclature – technically, the name does not translate in any other language. The letters that make up its name are an anagram of *Palinurus* the namesake of the family which was named after a helmsman in Virgil's *Aeneid*. A third *palinurid*, *Panilurus*, was named in the same way and the three names often cause confusion among crustacean biologists.

Class:	Malacostraca
Order:	Decapoda
Habitat:	Warm seas
Distribution:	Global
Time scale:	Cretaceous–Recent

Palinurina longipes

This little creature lived on the sea bottom, probably in waters so deep that it was too dark to see. *Palinurina longipes* was a type of lobster usually found preserved with its body curled up and its legs and other appendages pointing forwards. Scientists are unsure whether this was the usual position it adopted in life or whether it simply curled up like this when it died. *Palinurina longipes* was well adapted for life on the seabed. Its long antennae would have enabled it to feel its way around very easily in the darkness. *Palinurina longipes* grew to around 10cm (4in) long, although most of that length was made up by its feelers.

Order:	Decapoda
Family:	Palinuridae
Habitat:	Deep seabed
Distribution:	Europe
Time scale:	Jurassic

Harpactocarcinus punctulatus

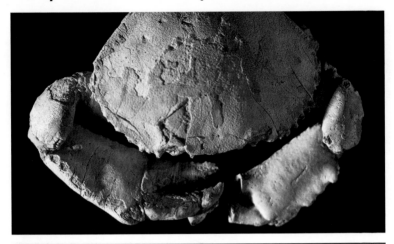

*H*arpactocarcinus was a large prehistoric crab with a wide, convex exoskeleton whose surface was almost totally smooth. There were regular notches along its front and side edges. *Harpactocarcinus* had particularly powerful pincers. Fossils of *Harpactocarcinus* have been found from the Eocene only, located in Italy, Spain and elsewhere in Europe, and also in North America. One specimen, discovered in the state of Texas, measured 7.9cm (3in) wide. *Harpactocarcinus punctulatus*, pictured here, was somewhat larger – between 8cm (3¼in) and 9cm (3½in) in width.

Class:	Malacostraca
Order:	Decapoda
Habitat:	Marine
Distribution:	Europe & USA
Time scale:	Eocene

Eryma modestiformis

This small crustacean was one of the first true species of lobster and dates back to the Jurassic Period (150 million years ago). *Eryma modestiformis* are also known as Solnhofen lobsters because some of the most spectacular fossils come from a band of limestone beds found in that region (situated between Nuremberg and Munich). During the Jurassic Period, this area lay on the edge of the Tethys Sea, whose fine, sandy beaches and silty lagoons provided ideal conditions for the formation of fossils. The claws of this specimen are missing, but the impressions they left in the rock are plainly visible, proving that they were preserved but subsequently lost.

Order:	Decapoda
Family:	Erymidae
Habitat:	Marine
Distribution:	Europe & South America
Time scale:	Jurassic

Thalassina *anomala*

*T**halassina anomala* is a small mud lobster that lives in Australia, burrowing in the tidal mudflats. It periodically moults like other crustaceans and buries the discarded shells in the bottom of its burrow before excavating a new living chamber. These buried shells can react with the surrounding sediments, resulting in the cementation of the surrounding mud – a type of mineralization. Since these mud lobsters moult many times during their life, and this mineralization can occur in perhaps less than a year, these remains may belong to an individual still burrowing away in a modern mudflat. Sometimes called 'instant fossils', these remains do not represent true fossils, as they are less than 10,000 years old. However, some of this type of *Thalassina* preservation are old enough to qualify as fossils.

Class:	Malacostraca
Order:	Decapoda
Habitat:	Tidal mudflats
Distribution:	Australia
Time scale:	Pleistocene–Recent

Palaxius isp. (coprolites)

Coprolites, or fossil faeces, are notoriously difficult to assign to an animal maker partly because excrement can look so similar from one animal to the next. This problem is compounded by the fact that, with fossils, we are usually dealing with long-extinct species never seen alive. This Late Cretaceous mass of callianassid, or ghost shrimp, coprolites, from the Mt. Laurel Formation in New Jersey, tells a different story. The gut anatomy of callianassids is such that it leaves a symmetrical and species-specific set of canals that runs down the length of each faecal pellet, just visible in some of the pellets here. Because of this specificity, modern ghost shrimp faeces can be used in ecological studies as reliable indicators of certain species. In the fossil record, these long-extinct species can be useful tools in stratigraphic studies.

Order:	Decapoda
Family:	Callianassidae
Habitat:	Marine muds& sands
Distribution:	Global
Time scale:	Cretaceous–Recent

Galatheid (moults and coprolites)

Because animals rarely die right after defecating, and in the same spot, palaeontologists usually have a tough time figuring out exactly which species made what faeces. There are some clues, like size, content, and geological context, that can get us close to an identification, but species resolution is very elusive. The Pliocene specimen shown here (in two views) comes from the Purisima Formation of Santa Cruz, California and preserves a very rare association. In the view on the left, the tiny, rod-shaped coprolites (trace fossils) of a crustacean are exposed at the surface. On the opposite side of the nodule are the moults of galatheid crabs (body fossils). Galatheids are the probable coprolite makers: the possibility of this being a chance relationship is incredibly small.

Order:	Decapoda
Family:	Galatheidae
Habitat:	Marine
Distribution:	Global
Time scale:	Jurassic–Recent

Ophiomorpha isp. (burrow)

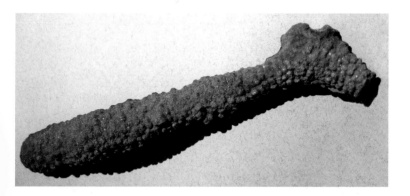

Trace fossils have their own separate taxonomy called 'ichnotaxonomy', meaning 'the taxonomy of traces'. It exists because of the many ambiguities that trace fossils present us with. *Ophiomorpha* is an ichnogenus of burrows belonging to marine ghost shrimp and their allies. This specimen comes from the Late Cretaceous Mount Laurel Formation of New Jersey and clearly shows the bumpy outer walls still seen in modern ghost-shrimp burrows. This texture is created as the shrimp adheres small balls of mud to the outer walls of vertical tunnels and the upper surface of lateral ones like this one. Finding the burrows proves the shrimp was present even in sediments that do not preserve remains of the crustaceans themselves. The presence of these burrows is also useful in palaeoenvironmental reconstructions.

Order:	Decapoda
Family:	Callianassidae
Habitat:	Marine muds and sands
Distribution:	Global
Time scale:	Cretaceous–Recent

Balanus sp.

*B*alanus is a barnacle, a marine crustacean that attaches itself to underwater surfaces, notably the bottoms of ships (scraping *Balanus* and other barnacles from the bottoms of ships was a regular task for sailors), but also to rocks along the shoreline. The specimens here had attached themselves to a dead scallop shell. *Balanus*, is shaped like a shortened cone constructed of six hollow-walled plates of calcium carbonate. These plates are mounted on a base which cements itself permanently to rock or other submerged marine surfaces. Through an opening called an operculum, *Balanus* collects food from the surrounding water with its six pairs of legs. That done, the operculum is closed by four valves.

Class:	Cirripedia
Family:	Balanidae
Habitat:	Marine surfaces
Distribution:	Global
Time scale:	Early Jurassic–Recent

Archimedes sp.

The name of this fossil, found mainly in North America and China, reflects its appearance. It resembles the Archimedes Screw, a device invented by the Ancient Greek mathematician Archimedes (287–212BC) to raise water by means of a three-dimensional spiral rotating inside a tube. Despite appearances, the bryozoan *Archimedes* performed a totally different function, as the axis of a colony of small organisms that arose by budding (dividing) from each other. (The organisms proved to be very fragile, and the axis was frequently the only part of the colony to become fossilized.) When complete, the colony consisted of twisted fronds resembling a net. The order name Fenestrata came from the window-like appearance of the apertures in the *Archimedes* colony.

Class:	Stenolaemata
Order:	Fenestrata
Habitat:	Marine
Distribution:	Europe & North America
Time scale:	Carboniferous

Fenestella sp.

Bryozoans like *Fenestella* lived in colonies. In appearance, *Fenestella* resembled a net and was formed by very thin, evenly spaced branches. These branches were joined by vertical dissepiments (plates). The individual members of the colony, or zooids, lay in two rows along the branches; their apertures (the opening in the margins of their shells) were situated on one side only. The dissepiments carried no zooids. There were spiny growths at the base of the colony, which sometimes carried sharp hooks. In common with other net-like bryozoans, *Fenestella* colonies obtained their food by filtering it from currents in the water moving through the holes in the colony. *Fenestella*, which was around 5cm (2in) in height, lived worldwide.

Class:	Stenolaemata
Order:	Fenestrata
Habitat:	Marine
Distribution:	Global
Time scale:	Ordovician–Permian

Bryozoan

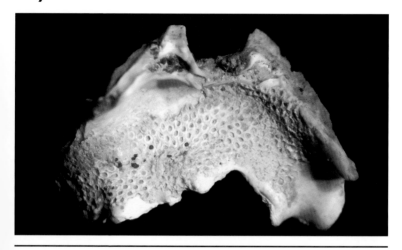

Bryozoans, or 'moss animals,' have a long history on Earth leading right up to the present but generally escape notice because of their small size and habits. They are colonial animals with a variety of forms living attached to underwater surfaces. The Late Cretaceous colony pictured here is an encrusting cheilostome bryozoan from New Jersey, USA. This type cements their individual chambers, each which housed an individual animal called a zooid, to smooth surfaces, in this case, the shell of the small oyster *Agerostrea*. Since the bryozoans in this specimen are covering the inside surface of the bivalve's shell, it is clear that they colonized a dead oyster. Weathering has worn away the outermost parts of the zooid's chambers leaving only their bases that outline the packed nature of the colony's individuals.

Class:	Gymnolaemata
Order:	Cheilostomata
Habitat:	Marine
Distribution:	Global
Time scale:	Cretaceous–Recent

Lingula sp.

Lingula has been called a 'living fossil' and it has remained virtually unchanged in the hundreds of millions of years it has lived on earth. Fossils show that *Lingula* were fairly small brachiopods with a thin shell and convex valves, and their modern representatives remain the same. In appearance, *Lingula* resembled a tongue; the Latin word 'ligula' or 'lingula' means 'little tongue'. The pedicle valve on the underside of this brachiopod had a groove to support the pedicle. *Lingula* had no teeth or sockets for teeth. Its habitat was probably the same as its descendants use today – in between tidal areas in vertical burrows. More than 3000 brachiopod species have been described by their fossils, and they are the most prolific fossils to be found in Palaeozoic rocks.

Class:	Lingulata
Order:	Lingulida
Habitat:	Marine muds
Distribution:	Global
Time scale:	Cambrian–Recent

Orbiculoidea sp.

The valves of the inarticulate – that is, hingeless – brachiopod *Orbiculoidea* were held together solely by muscle power. Likewise, *Orbiculoidea* opened its valves by using its muscles and ligaments. One valve was flat, enabling it to adhere to a surface such as a rock, or rocky sea bed. *Orbiculoidea's* thin-walled conical shell was formed from fibrous chitin and phosphate. The umbo was situated off-centre in the lower valve. In some specimens, the slit in the stalk behind the umbo could reach the edge of the valve. On its surface, *Orbiculoidea's* shell carried fine concentric growth rings. *Orbiculouidea's* fossils have been found all over the world. This brachiopod was relatively small, with a valve measuring around 7mm (¼in) compared to over 1cm (⅜in) for others of the same genus.

Class:	Lingulata
Order:	Acrotretida
Habitat:	Marine
Distribution:	Global
Time scale:	Ordovician–Cretaceous

Pygites diphyoides

Brachiopods are an extremely ancient lineage with their origins all the way back at the beginning of the Phanerozoic (around 540 million years ago). In fact, the earliest known brachiopods group still exists today, apparently unchanged in half a billion years. They persist, with a much reduced diversity and range, but their history shows a bewildering variety of form and habitat. The central perforation in the shell of *Pygites* is peculiar to the members of a small short-lived group called pygopids. *Pygites*, like many brachiopods, was an epifaunal suspension feeder, meaning it lived its adult life on the sea floor, filtering plankton and other suspended particles of food from the water. The perforation might represent an adaptation to deep, nutrient-poor waters.

Order :	Terebratulida
Family:	Pygopidae
Habitat:	Deep marine
Distribution:	Europe
Time scale:	Late Jurassic–Early Cretaceous

Paraspirifer bownockeri (pyritized)

P*araspirifer* was typical of its order, being noticeably broader than it was long. The two valves comprising its wide shell bore narrow ribs. *Paraspirifer* lived during the Devonian Era, the time when the first amphibians appeared on Earth. The picture above shows a *Paraspirifer* fossil, measuring 5cm (2in) from the species *Paraspirifer bownockeri*. This particular specimen, which dated from the Middle Devonian, was found at Sylvania, Ohio, in the United States. It is beautifully preserved in the minerla pyrite, which is common for the *Paraspirifer* fossils of the area.

Class:	Articulata
Order:	Spiriferida
Habitat:	Marine
Distribution:	Global
Time scale:	Lower–Middle Devonian

Spiriferina sp.

Spiriferina is one of more than three thousand identified brachiopods. Brachiopod fossils predominated among those found in Palaeozoic rocks. The shell of *Spiriferina*, 15cm (6in) long and shaped somewhere between a triangle and a five-sided pentagon, was composed of two convex valves – a brachial (upper) valve and a prominent pedicle (lower) valve, which attached *Spriferina* to a surface. The pedicle valve carried umbones (beak-like projections), and an interarea (a flat area between the shell hinge and the beaks) was strongly developed. Both *Spiriferina's* valves carried large round ribs and some had fine grooves. *Spiriferina*, which has been found all over the world, 'sat' with its umbones on the muddy sediments where it lived.

Class:	Articulata
Order:	Spiriferida
Habitat:	Marine
Distribution:	Global
Time scale:	Silurian–Middle Jurassic

Leptaena sp.

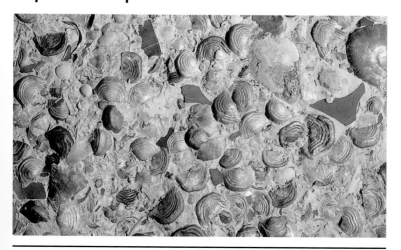

*L*eptaena's shell, which was semi-circular, had a wavy orientation and a straight hinge line. There was a slightly convex valve to *Leptaena's* pedicle, which attached it to the sea floor on which it lived. During life, *Leptaena's* brachial valve, which comprised half of its shell, was almost completely buried in the sediment. At the front, both of *Leptaena's* valves bent almost at right angles to each other, while the rest of its front was flat. The rugae, the wrinkles that decorated the surface of *Leptaena's* shell, may have developed in concentric form as a way of enabling the shell to remain stable in soft surfaces.

Class:	Articulata
Order:	Strophomenida
Habitat:	Marine
Distribution:	Global
Time scale:	Carboniferous

Mucrospirifer sp.

Mucrospirifer was a spectacular spiriferid with a shell that was formed in the shape of outstretched 'wings'. It was relatively large, measuring 2.5cm (1in) in length. The valves that formed *Mucrospirifer's* shell reached ultimate width along the hinge line. In fact, in ancient China, spiriferid fossils like these were called 'stone swallows' and were supposed to have magical properties. *Mucrospirifer's* shell was decorated with ribs, which ended in overlapping growth layers at the edges. *Mucrospirifer* lived mainly in Europe and North America.

Class:	Articulata
Order:	Spiriferida
Habitat:	Marine
Distribution:	Global
Time scale:	Devonian Period

Terebratula sp.

Terebratula is a large brachiopod with a long elliptical shell containing convex valves. The umbo (where the shell first started to form) is short, but big and lightly curved in. There was a thick fleshy pedicle (the appendage used by brachiopods to attach themselves to rocks and other hard, stable surfaces on the seabed). Inside the shell, *Terebratula* had prominent teeth and sockets and a short septum (thin dividing wall) on its brachial valve (one of two that form the shell). They were covered in fine growth lines, becoming most noticeable near the margins of the shell. *Terebratula*, which lived mainly in Europe, was informally known as the 'lamp shell' for its resemblance to an old-fashioned oil lamp.

Class:	Articulata
Order:	Terebratulida
Habitat:	Marine
Distribution:	Global
Time scale:	Devonian–Recent

193

Strophomena sp.

Strophomenid brachiopods are the most abundant brachiopods of the Ordovician and the largest order of brachiopods, with about 400 genera. One of them is *Strophomena*, seen here as mostly disarticulated valves. The outside surface of the shells have a delicate radiating ornament and the inner surfaces show the complex muscle attachments. *Strophomena's* two valves nested somewhat, one within the other: one was convex and the other concave. Many scientists think these brachiopods rested on the sea floor convex side down. But others think the opposite, pointing out that the convex valves of the living brachiopods were the ones most colonized by encrusting epibionts (organisms that live attached to others). Strength was added to this argument by the fact that little of the encrustation occurred after death.

Order:	Articulata
Family:	Strophomenida
Habitat:	Marine
Distribution:	Global
Time scale:	Ordovician – Devonian

Pentamerus sp.

*P*entamerus was relatively large as brachiopods went – around 4.5cm (1¾in) in length. Fine growth lines decorated the outside of the shell, which was otherwise quite smooth. Pentamerus was able to attach itself to rocks, but adults made up for their inability to do so by turning umbo-down onto muddy or silty substrate that held them in place. *Pentamerus* was often found in shallow-water limestone areas, so much so that they frequently formed banks of shells. The Silurian *Pentamerus laeuis* lived in closely packed colonies such that whole surfaces of their fossils are found on the undersides of sandstone beds.

Class:	Articulata
Order:	Pentamerida
Habitat:	Marine
Distribution:	Northern Hemisphere
Time scale:	Silurian–Devonian

Productus sp.

Productus fossils have been discovered in Russian Asia, North Africa and China, with a few specimens in North America tentatively named as the same genus. *Productus* had a calcareous (chalky) shell. There was a convex valve on its venter (underside) and a hinge line between its two valves that overlapped. The valve on the upper surface of *Productus* was flat, giving the impression of a long jar with a lid. This arrangement was typical of the Strophomenids, which generally featured an unmatched pair of one concave and one convex valve. The surface of *Productus* was ornamented by radiating ribs and a scattering of spines. There were two rows of spines near the hinge-line on the underside.

Class:	Articulata
Order:	Strophomenida
Habitat:	Marine
Distribution:	Northern Hemisphere
Time scale:	Silurian–Permian

Tullimonstrum gregarium

A round half of all US states have a designated 'state fossil', but Illinois is one of the very few that has its own unique fossil. *Tullimonstrum,* or 'Tully's Monster', is found in abundance in Illinois' Mazon Creek fossil beds and no where else on Earth. Named after amateur collector Francis Tully, who found the first example in 1958, this soft-bodied marine creature had a long worm-like body. At one end was a pointed snout; at the other, a finned tail. Eyes on stalks projected outwards from the creature's 'head'. Inside its jaw were eight small razor-sharp teeth, probably used to pierce the bodies of prey and suck out juices. So far *Tullimonstrum* has not been linked to any known species.

Order:	Unclassified
Family:	Unclassified
Habitat:	Lowland swamps
Distribution:	Illinois, USA
Time scale:	Pennsylvanian

Achistrum sp.

Achistrum fossils are small and not very spectacular, but they are important evidence of the ancient existence of a group of animals still alive today. *Achistrum* was a type of sea cucumber that lived on the ocean floor more than 300 million years ago. Just like sea cucumbers today, it fed by crawling over the seabed and swallowing the sand and other sediment as it went – a bit like an earthworm, but working on the sea floor instead of within the soil. *Achistrum* digested all of the tiny bits of organic matter mixed up with the sediment it swallowed, then excreted the rest. Sea cucumbers are related to sea urchins and starfish, which had also evolved by the time the Carboniferous Period began.

Class:	Holothuroidea
Order:	Apodida
Habitat:	Ocean floor
Distribution:	Northern Hemisphere
Time scale:	Carboniferous–Jurassic

Furcaster sp.

This creature belonged to a now-extinct group of echinoderms closely related to modern brittle stars. Like them, it moved by 'walking' over the seabed with its five limbs, meaning that, compared with most other echinoderms, it could move relatively fast. It may have even been able to swim, as a few brittle stars can do today. *Furcaster* was an extremely common inhabitant of the Tethys Sea. Large numbers of its fossils have been found in Germany in particular, in rocks laid down during the Devonian. It probably fed on the bodies of dead animals that sank down on to the sea floor. Its legs were each about 8cm (3¼in) long, but the tiny body at their centre was less than 1cm (½in) across.

Order:	Oegophiuroidea
Family:	Protasteridae
Habitat:	Seabed
Distribution:	Europe and the Americas
Time scale:	Devonian

Helianthaster sp.

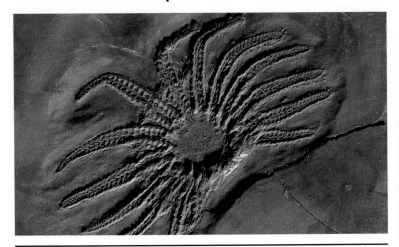

Helianthaster was an ancient sea star with 16 limbs, which stretched out from its body like rays from the sun (*Helianthaster* actually means 'sun star' in Latin). Like all sea stars, it lived on the ocean floor, moving slowly along by means of hundreds of tiny tube feet on its underside. *Helianthaster* lived around 390 million years ago in the Tethys Sea, which then covered Germany and much of the rest of Europe. It probably fed on shellfish and other creatures that lived fixed to the sea floor, as most starfish do today. Only a few examples of *Helianthaster* have ever been found, which suggests that it was rare even when it was alive.

Order:	Hemizonida
Family:	Helianthasteridae
Habitat:	Seabed
Distribution:	Germany
Time scale:	Early Devonian

Budenbachia beneckei

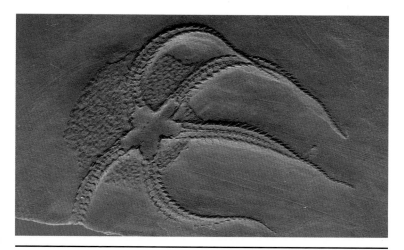

This fossil ophiuroid, or brittle star, comes from the Hunsrück Slate lagerstätten of Bundenbach and Gemünden, Germany. A lagerstätten is a palaeontological site that contains extraordinary preservation or abundance of its fossils, or both. In the case of the Hunsrück, the preservation of the organisms is what makes it special. Many of the fossils are pyritized and show up well on X-rays, illuminating the delicate structures still hidden in the rocks. This deposit is one of the few that preserves the fragile antennae, gills and legs of trilobites. Another thing that is commonly shown in these fossils is the direction of the current at the time of burial, and can be discerned clearly in this *Budenbachia* with all of its arms trailing in the same direction.

Order:	Oegophiuroidea
Family:	Protasteridae
Habitat:	Marine
Distribution:	Germany
Time scale:	Devonian

Asteriacites isp. (resting traces)

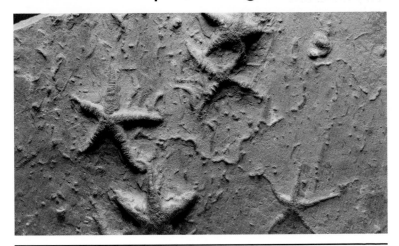

*A*steriacites is the name given to trace fossils formed by five-armed sea stars. (A trace fossil is a fossil that shows where an organism has been or what it has done, rather than the body or remains of the organism itself.) *Asteriacites* formed on muddy sea floors around sea stars at rest. When the starfish moved away, the shapes of their bodies were left in the soft mud. Although they were soon covered up by more sediment, they remain, millions of years later, as impressions in the rock. *Asteriacites* are often found together in large numbers and show just how abundant starfish were in ancient seas.

Sub–Phylum:	Echinodermata
Class:	Asteroidea
Habitat:	Ocean floor
Distribution:	Global
Time scale:	Ordovician–Recent

Clypeaster sp.

Clypeaster, which measured around 12cm (4¾in) across, had a large, thick, round shell, which was roughly five-sided in shape. When alive, the shell was covered in short spines. *Clypeaster's* tube feet were enlarged on the top surface, forming petals bearing numerous long pores. There was a small circular mouth lying slightly below the centre of the surrounding jaws. The jaws of sand dollars and sea urchins form a structure known as 'Aristotle's Lantern' because of its appearance. *Clypeaster* lived right across the world, spending its time submerged in sand in shallow water. It fed on the organic and other particles in the sediment that settled around it.

Class:	Echinoidea
Order:	Clypeasteroida
Habitat:	Marine sands
Distribution:	Global
Time scale:	Eocene–Recent

Clypeus sp.

Clypeus had a large rigid shell, shaped like a flattish disc, with one flat side, the bottom, where its mouth was located. The plated area around its tube feet was clearly visible on top of the shell. The markings on the shell were shaped like petals, which met in the centre and spread out to the edges of the shell. A deep groove extending to the back of the shell contained the anus, which was sited off-centre. At the front of the shell, the tooth-like appendages around the mouth formed a five-sided shape. *Clypeus* could grow to a diameter of 10cm (4in) and foraged for its food on the sea bed. It lived worldwide, but its main habitats were in Europe and Africa.

Class:	Echinoidea
Order:	Cassiduloida
Habitat:	Marine
Distribution:	Europe
Time scale:	Jurassic

Hemipneustes sp.

*H*emipneustes was a sea urchin with a slightly elliptical shell that could grow to a height of about 10cm (4in). The shell surface was smooth and the plates themselves took the form of elongated petals. The peristoma (the opening in the shell for the mouth) was shaped like a half-moon. *Hemipneustes* lived in the seas around Europe, Africa and Asia. It partly buried itself in the sand on the seabed, which can be surmised from the condition in which many are found: the buried lower half usually survived, but to judge from the number of 'headless' *Hemipneustes* found, the top half fell easy victim to predators. Limburg in Belgium has proved to be a particularly fruitful source of *Hemipneustes* fossils.

Class:	Echinoidea
Order:	Holasteroida
Habitat:	Marine
Distribution:	Europe
Time scale:	Cretaceous

Micraster sp.

Micraster is one of the urchins known as 'heart urchins'. It had a sharp point at the back and a distinct notch at the front. The high domed top was decorated with fine tubercles (raised bumps), which supported tiny spines in life, and the top was incised with five petals, some of which reached part-way across and others, to the edge. *Micraster*, which was one of the more advanced forms of its type, lived on the chalk sea floor of the Cretaceous, buried in the soft mud. The notch at the front enabled it to channel a stream of mud into its mouth. *Micraster* lived all over the world for the 55 million years it existed on Earth. On average, *Micraster* measured around 5cm (2in) in diameter.

Class:	Echinoidea
Order:	Spatangoida
Habitat:	Marine
Distribution:	Europe, Africa, Antarctica
Time scale:	Cretaceous–Paleocene

Phymosoma sp.

The hard external covering of *Phymosoma*'s shell was a flattened sphere, with very broad areas in the upper central disc, or apical, that were characteristic of sea urchins, like this one. Likewise, the membrane around *Phymosoma*'s mouth was broad. The plates around its feet and the parts between the plates measured roughly the same size and displayed smooth tapering spines that were either cylindrical or flat. It is quite common for spines to be found on their own. *Phymosoma*'s habitat was the bed of the chalk seas. It scraped hard surfaces in order to obtain its food, which came in the form of algae, sponges or other soft organisms.

Class:	Echinoidea
Order:	Phymosomtoida
Habitat:	Marine
Distribution:	Europe
Time scale:	Jurassic–Eocene

Lenticidaris utahensis

*L*enticidaris utahensis was a typical cidarid sea urchin. What makes it special, in part, was its discovery. Until this taxon was described in 1968, no whole fossil urchins had ever been found in the Triassic of North America. Two Triassic species were previously described from California, but they were both based on fragmentary specimens. When Mr Afton Fawcett uncovered more than 200 specimens of what would become known as *Lenticidaris* in the Virgin Formation of Utah, it was a very significant discovery. Aside from the gap in the fossil record that it helped to fill, the accumulation was the largest Triassic amassing of sea urchins yet known. The species is also unique itself in having a highly flexible upper surface. The physiological importance of this feature, though exceptional, is currently unknown.

Order :	Cidaroida
Family:	Miocidaridae
Habitat:	Marine
Distribution:	USA
Time scale:	Early Triassic

Cidaris coronata

The sea urchins are a diverse group of echinoderms. Within the group are the very successful cidarids, or regular echinoids – urchins with prominent spines attached to their sub-spherical test (shell). The attachment points of the spines form a striking pattern on the test, easily seen on this fossil test of this Jurassic *Cidaris coronata*. Cidarids make their living scavenging food off the sediment surface or rocks, which calls for powerful jaws and a life exposed on the sea floor. Because of this and their slow movement, they need a strong defence: large and sometimes elaborate spines are the characteristic armament of this group. The 'regular' part of 'regular echinoids' refers to the radial symmetry of their body plan, which allows travel in any direction in accordance with the animal's roaming mode of life.

Class:	Echinoidea
Order:	Cidaroida
Habitat:	Marine
Distribution:	Global
Time scale:	Jurassic

Encope sp.

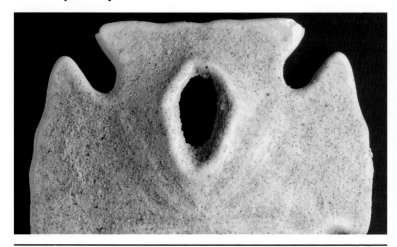

*E*ncope is popularly known as a sand dollar because of its very flat, thin shape. Its diameter measured around 9cm (3½in). There were five 'petals' in the centre, which were more prominent in some species – such as *Clypeaster* – than in others: in *Encope*, the petals are very faintly outlined. The upper surface of *Encope* was perforated, close to the edge, by five oval lunules (holes or notches) situated along the outer edge. There was another, rather larger, lunule towards the back. These helped *Encope* to feed off the fine sediment of its habitat. Fossils of *Encope* have been found in both North and South America, where it lived in shallow warm seas, buried in the sand.

Class:	Echinoidea
Order:	Clypeasteroida
Habitat:	Sandy seabed
Distribution:	Tropical North & Central America
Time scale:	Miocene–Recent

Dendraster sp.

This flattened animal was what we know today as a sand dollar, a close relative of sea urchins and starfish. *Dendraster* lived by filter feeding, catching tiny animals and other particles of food that floated past with sticky mucus secreted by special glands. In life, it often positioned its body vertically (like a coin standing on its edge) to increase its chances of catching passing food. Part of its rim was buried in the sand to stop it falling over. Like starfish, *Dendraster* had tube feet, which enabled it to move slowly across the sea floor and lift itself into an upright position. It also had a covering of short protective spines.

Order:	Clypeasteroida
Family:	Dendrasteridae
Habitat:	Sandy seabed
Distribution:	North & South America
Time scale:	Pliocene–Recent

Isorophus cincinnatiensis

In 2002, Cincinnati citizens voted to make their most famous 'resident' – *Isorophus cincinnatiensis* – the city's official fossil. This distant relative of the starfish lived in the region 450 million years ago, when the area was a warm tropical sea. These small, disc-shaped creatures have a simple outer body (theca) made up of scale-like plates. A ring of muscle (the peripheral rim) enables the echinoderm to attach itself to rocks on the sea bed while its five arms gather food. Today, fossilized Cincinnati echinoderms are so well known that the period they date from is often called 'the Cincinnati Age'. Approximately 10 species of edrioasteroids (the oldest group of echinoderms) have so far been found in the state.

Class:	Edrioasteroidea
Order:	Isorophida
Habitat:	Shallow tropical seas
Distribution:	Ohio and Illinois
Time scale:	Ordovician

Holocystites scutellatus

The echinoderms are one of the strangest groups of marine invertebrates and are exemplified by the modern urchins, sea stars, brittle stars, sea biscuits, sand dollars, sea cucumbers and crinoids. Cystoids were a very primitive group of echinoderms related to the crinoids and blastoids. The calyx (body) exhibits body plates that are irregular in arrangement, whereas crinoids, and especially blastoids, are noted for their more regular plate arrangements. Cystoid arms were irregular, delicate and rarely preserved. Like their close relatives, they were normally stalked animals, living their filter-feeding lives attached to rocks or other stationary objects, but some genera occur both with and without stalks. Cystoid fossils can be found in rocks from the Ordovician Period until the end of the Mississippian Period.

Phylum:	Echinodermata
Class:	Diploporita
Habitat:	Marine
Distribution:	Eastern North America
Time scale:	Silurian (*Holoycistites scutellatus*)

Echinosphaerites aurantium

*E*chinosphaerites aurantium* was an ancient relative of today's sea urchins and starfish. Unlike them, it lived attached to the seabed, held on by a thin fleshy stalk (which rarely fossilized). *Echinosphaerites aurantium* fossils are sometimes known as 'crystal apples' because of their unusual shape. They are common in Ordovician rocks, which formed between 505 and 438 million years ago. Scientists are unsure exactly how *Echinosphaerites aurantium* lived, but it probably fed on tiny animals and particles of food that were swept past it by the current. The specimens seen here are broken open, revealing the crystal-lined interiors of their hollow bodies. These geodized fossils do not reflect the interior structure of the actual animal.

Class:	Rhombifera
Family :	Echinosphaeritidae
Habitat:	Marine
Distribution:	Northern Hemisphere
Time scale:	Ordovician

Crotalocrinus sp.

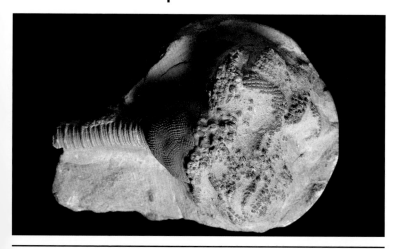

*C*rotalocrinus, which like other crinoids was popularly known as a sea lily, had a semi-circular or round cup at the top of its stem, which consisted of three rings of plates. One of these rings was an anal plate, which was joined to one of the plates at the base of the crinoid. The top of the cup was made up of five fused mouth plates of unequal size. When closed up, the cup grew to a height of 3.5cm (1⅜in). *Crotalocrinus* fossils have been found all over the world in the strata laid down during the Silurian era.

Class:	Crinoidea
Phylum:	Echinodermata
Habitat:	Marine
Distribution:	England
Time scale:	Silurian

Crinoid

The crinoids have been on earth for most of the Phanerozoic. In some rocks, especially those of the Mississippian, it is clear that they formed a major part of the world's marine faunas and exhibited a tremendous diversity of species. In fact, they were so common that this time period is often referred to as the Age of Crinoids. Many rocks of the time are composed nearly entirely of the fragmented stems of crinoids: 'crinoidal limestones' are calcium carbonate rocks in which crinoid parts predominate. Several pieces of crinoid stems can be seen in this specimen, preserved as natural moulds. The actual material of the stems dissolved away, leaving a cavity. Following the cavity's central axis is the sediment that filled the hollow cavity of the crinoid stem.

Class:	Crinoidea
Phylum:	Echinodermata
Habitat:	Marine
Distribution:	Global
Time scale:	Ordovician–Recent

Cyathocrinites sp.

Crinoids are popularly known as sea lilies. They possessed a large calcite skeleton. Vast numbers of them lived in the prehistoric seas, and in time, their remains formed extensive deposits of limestone. *Cyathocrinites* consisted of branched 'arms' shaped like columns growing from a rounded stem. The 'cup' shaped by the branches was around 8mm (¼in) in diameter and the the branches themselves formed a 'crown'. The conical cup positioned just above the stem had rounded plates, and there was a plated anal tube on the opposite side. *Cyathocrinites* arose in the Silurian Period, when the first land plants appeared. Found mainly in Europe, their habitat consisted of shallow, calm waters where they could find ample food.

Class:	Crinoidea
Order:	Cladida
Habitat:	Marine
Distribution:	England & North America
Time scale:	Silurian–Permian

Encrinus lilliformis

The calcite skeletons of crinoids like *Encrinus lilliformis* served as the raw material for the building of extensive limestone deposits. *Encrinus*, which lived in the shallow seas of Triassic Europe, had a long cylindrical stem with enlarged columnals set at regular intervals. The stem led to a crown that comprised ten short but very strong 'arms'. To trap plankton (the microscopic organisms on which it fed), *Encrinus* opened out its arms into the shape of a cup or fan, which was around 2.5cm (1in) in diameter. The plankton trapped inside were moved to *Encrinus'* mouth along grooves inside its 'arms'. When *Encrinus* was in danger, the arms closed into a tight mass – ready, perhaps, to fend off an enemy.

Class:	Crinoidea
Order:	Encrinida
Habitat:	Marine
Distribution:	Europe, Asia, New Zealand
Time scale:	Triassic

Saccocoma pectinata

*S*accocoma was a crinoid without the long stem that characterized its class. Nevertheless, *Saccocoma* had the typical cup, in this case, a small globular cup about 2cm (¾in) in diameter, made up of the usual ten 'arms'. Small extensions like wings sprouted from the lower sections of its arms and there were thin side-branches growing from the upper sections. The leaf shapes at the end of the tentacles may have helped *Saccocoma* to float easily in water. Most of the fossils discovered in the Solnhofen limestone in Bavaria, southern Germany, have been *Saccocoma* fossils and many of the most complete specimens have been found there.

Class:	Crinoidea
Order:	Roveacrinida
Habitat:	Lagoons
Distribution:	Europe, North Africa, Cuba, North America
Time scale:	Late Jurassic

Pentacrinites sp.

The crinoids are one of those special groups that were known as fossils before they were found living today. This is mainly because they have such an extraordinarily rich fossil record and because they currently live in mainly deep water environments – and thus have been in areas relatively recently discovered – with much reduced diversity. Their common name is 'sea lilies', and most assume that this is in reference to their flower-like body form with its accompanying attachment to the sea floor. In truth, their comparison to lilies comes from the flower-like form of columnals like those pictured here from *Pentacrinites*, which clearly exhibit the fivefold symmetry common among the echinoderms. The stem of crinoids is a series of stacked calcite elements, which often disarticulate after death.

Class:	Crinoidea
Order:	Sagenocrinida
Habitat:	Marine
Distribution:	Global
Time scale:	Triassic–Eocene

Pentremites sp.

The thecae of *Pentremites* are beautiful little fossils. Reminiscent of plant seed pods (sometimes erroneously referred to as 'fossil hickory nuts') they actually represent the main part of the body of an animal closely related to sea stars and crinoids. These echinoderms had thin stems that attached to a hard surface and tiny arms that extended upwards from the five-pointed ambulacral system at the top of the theca, with which they collected their planktonic food. The slender nature of the stem, plus the fact that its individual columnals tended to disarticulate after death, makes whole specimens of *Pentremites*, in fact any blastoid, very rare. But the highly durable thecae are well represented in the fossil record. The blastoids existed from the Middle Ordovician to the Permian.

Class:	Blastoidea
Order:	Spiraculata
Habitat:	Marine
Distribution:	North America
Time scale:	Carboniferous

Homalozoan

Echinoderms have a long history on Earth and exhibit an extreme diversity of form and life habit right up to the present. Their unusual forms, especially many of the long-extinct forms, have intrigued and baffled scientists for centuries. One of the most bizarre groups is the Paleozoic homalozoans. They were mobile echinoderms with thin tails and calcite exoskeletons, and at least some seemed to have propelled themselves through the muddy sea bottom with repeated hooking of their tails. What is most vexing about them is their apparent mixture of echinoderm and chordate features, prompting some scientists to place them as ancestral to both groups. Others consider them to be an odd branch of the echinoderm tree that merely retains some of the ancestral features of the echinoderm/chordate ancestor.

Order :	n/a
Family:	n/a
Habitat:	Marine
Distribution:	Global
Time scale:	Cambrian–Carboniferous

Dictyonema sp.

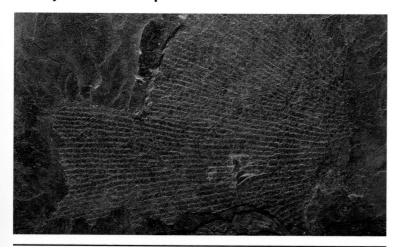

Dictyonema was a member of the graptolite group Dendroidea, a group of sessile, colonial, filter-feeding animals with a conical net-like colony attached to a surface by a thread called the nema. Graptolites are classified as hemichordates and may be closely related to ancestor of all chordates, and thus, vertebrates. More than once, living organisms have been declared living graptolites, most recently, in 1993 Professor Noel P. Dilly described the extant pterobranch hemichordate *Cephalodiscus graptolitoides* as a living graptolite, but a scientific consensus has not yet been reached. *Dictyonema* the graptolite should not be confused with the *Dictyonema* the lichen. The reason these two organisms can have the same genus is because one set of naming rules for animals and another for plants and fungi.

Class:	Graptolithina
Order:	Dendroidea
Habitat:	Marine
Distribution:	Global
Time scale:	Ordovician–Devonian

Monograptus sp.

Monograptus, another colonial creature, supported its individual members on a single branch. This single branch, however, could take different forms – some were straight, some gently curved and other were totally spiral. The spiral branch, or 'stipe', had a typical diameter of 20cm (8in). Graptolites of the *Monograptus* genus were a peripatetic set of creatures, floating freely in earth's open seas, and during the long span of time they existed, they spread all around the world.

Class:	Graptolithina
Order:	Graptoloidea
Habitat:	Marine
Distribution:	Global
Time scale:	Ordovician–Devonian

Megaderaion sinemuriense

Hemichordates such as *Megaderaion sinemuriense* are important because these tiny marine 'worms' allow naturalists to understand more about the development of the vertebrae or backbone. (Hemichordata means 'half-chordate': Chordata is the group that includes vertebrates.) There are three known classes, which are called 'sawblades' and 'acorn worms'. *Megaderaion sinemuriense*, which was discovered in Osteno, Italy in 1981, is believed to be a species of acorn worm. This fossil shows an imprint of a soft-bodied invertebrate, measuring 2cm (¾in) long. Look closely, and you will notice that its body is split into three distinct divisions – preoral lobe, collar and trunk – which is a common design in hemichordates. To date, clear scientific classification of hemichordates has been difficult due to the poor fossil record.

Phylum:	Hemichordata
Class:	Enteropneusta
Habitat:	Marine
Distribution:	Osteno, Italy
Time scale:	Jurassic

Streptognathodus isolatus

The conodonts were fossil animals that covered the whole Palaeozoic and a bit of the Triassic. They are almost exclusively represented by tooth-like elements of variable form, which are fantastically abundant microfossils (the ones pictured here are 1mm long) and can be found in nearly any marine sediments of the right age. Despite their profusion, they have defied classification since their initial description in 1856. They have been considered fish teeth, the scraping elements of snail tongues, spines of trilobites or crustaceans, worm jaws, and parts of a variety of molluscs and worm-like animals. In the 1970s, a fossil animal was found with the tooth-like elements inside, but it turned out to be a conodont predator. Only recently have true whole conodonts been found and they strongly suggest a vertebrate affinity.

Order:	Conodontophorida
Family :	n/a
Habitat:	Marine
Distribution:	Kansas, USA
Time scale:	Late Pennsylvanian or Early Permian

Drepanaspis sp.

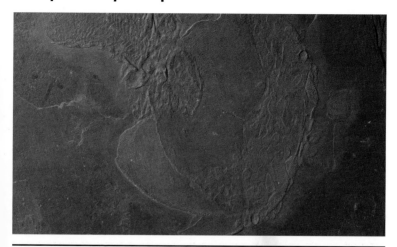

*D*repanaspis was a bizarre-looking armoured fish, which lived in the sea that covered Devonian Europe. Nothing quite like it exists in the oceans today – in fact, all the members of its order had died out before the first dinosaurs appeared. *Drepanaspis* had large bony plates on its head and back. The scales between these plates were also extremely tough, to protect the fish from predators. Because it had such a weighty, flattened body, scientists think that *Drepanaspis* lived on the seabed, where it fed on shellfish. Instead of teeth, its mouth contained large, flat, bony plates, which it used to crush its prey, exposing the flesh inside.

Class:	Pteraspidomorphi
Order:	Pteraspidomorphes
Habitat:	Marine
Distribution:	Europe
Time scale:	Devonian

Pterichthyodes sp.

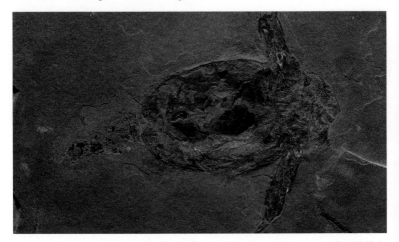

*P*terichthyodes* was a small fish, around 15cm (6in) long, and as a placoderm, it was something of a curiosity. It had a covering of armour, which possibly developed to help it crawl safely along the bed of the shallow lakes where it lived in what is now the United Kingdom. Its armour consisted of appendages with joints that were articulated with its body. *Pterichthyodes* had a somewhat small head, which was enclosed in a bony shield. The eyes were at the top of the shield and the jaws, at the bottom, might have been used to shovel food into the fish's mouth. *Pterichthyodes'* trunk was protected by a large shield, flat along the underside, which was made up of overlapping bony plates.

Class:	Placodermi
Order:	Antiarcha
Habitat:	Marine
Distribution:	Europe
Time scale:	Devonian

Colobodus sp.

This fish lived along the shores of the Tethys Sea, which covered most of Europe when the first dinosaurs were roaming the earth. It fed on shellfish that lived along the sea bottom, lifting them up in its mouth and then crushing them with its flat plate-like teeth to expose the soft flesh inside. The coastal waters *Colobodus* inhabited varied from warm and tropical in the south of its range to cooler and more temperate, like those that lap the shores of Europe today. *Colobodus* was a primitive fish compared with most living species. Its body was covered with large overlapping scales, which would have given its skin a much rougher texture than that of most fish alive today.

Class:	Osteichthyes
Order:	Perleidiformes
Habitat:	Coastal marine
Distribution:	Europe
Time scale:	Triassic

Xenacanthus sp. (teeth)

One look at these teeth and it is clear that they came from a shark. *Xenacanthus*, however, was unusual in that it lived in fresh water rather than the sea. The first species of *Xenacanthus* appeared long before there were dinosaurs on Earth. When it was alive, sharks were among the most common large creatures on the planet. Like most prehistoric sharks, *Xenacanthus* is known mainly from its teeth and a few spines. Sharks have skeletons made of cartilage rather than bone and, this being softer, it rarely fossilizes. Teeth, on the other hand, fossilize well. The teeth of *Xenacanthus* show the triangular shape and sharp edges typical of sharks.

Class:	Chondrichthyes
Order:	Xenacanthiformes
Habitat:	Lakes & rivers
Distribution:	Europe and the USA
Time scale:	Carboniferous–Triassic

Listracanthus sp. (fin spines)

*L*istracanthus was a very common type of shark that lived in the oceans just before and at around the same time as the first dinosaurs were appearing on the land. It is known only from spines, like the ones shown here, and rare patches of fossilized skin. The spines are large, which suggests that *Listracanthus* itself was a very large shark. *Listracanthus* is unusual in that it is one of the few creatures found both in Paleozoic and Mesozoic rocks. This means that it must have survived the massive extinction event that occurred between these two eras around 245 million years ago, when most other species on earth were wiped out.

Order:	Cladodontida
Family:	Uncertain
Habitat:	Marine
Distribution:	Global
Time scale:	Paleozoic–Mesozoic

Helicoprion sp. (tooth whorl)

It is clear from the microstructure of the teeth and certain details of their mode of replacement that *Helicoprion* was a shark or, at least, a very close relative of the group. Beyond that, it is very hard to determine what this animal looked like or how its jaws functioned. Its relationships within cartilaginous fish are also very hard to determine. This specimen from the Waterloo Phosphate Mine, near Montpelier, Idaho, USA shows beautiful spiral of teeth – all that has ever been found for this genus. It is this lack of associated body parts that makes knowledge of this creature so elusive. Until a complete skeleton or soft part anatomy is found with the teeth in place, *Helicoprion* will probably remain very mysterious.

Order:	Eugeneodontida
Family:	Agassizodontidae
Habitat:	Marine
Distribution:	Northwestern USA
Time scale:	Permian

Carcharocles megalodon

*C*archarocles megalodon was an enormous prehistoric shark measuring at least 12m (40ft) long, around three times as large as its modern relative the Great White Shark. *Carcharocles'* teeth could approach 20cm (8in) long and each individual had dozens of active teeth in its jaws at any given time, with hundreds more waiting for use within the jaws, like other sharks. Hard, thickly enamelled teeth are normally the only fossils left behind by *Carcharocles*, since it had cartilage instead of bones, which did not fossilize well. *Carcharocles* weighed up to 30 tonnes (33 tons) and its fossils have been found over a wide area – in Europe, India, the South Pacific and in both North and South America.

Class:	Chondrichthyes
Order:	Lamniformes
Habitat:	Marine
Distribution:	Global
Time scale:	Eocene–Pleistocene

Shark (vertebra)

Shark teeth are certainly one of the most common vertebrate fossils. Not only are tens of thousands produced, and dropped, by each shark during its lifetime, but sharks themselves are, and have been, extremely common in the seas of the world. Additionally, the teeth are made of materials that are very durable and stand up well to the rigours of becoming a fossil. The rest of a shark's skeleton is composed of cartilage, which decays rapidly with much less chance of being preserved. This has led many to state that teeth are the only parts of sharks found as fossils. However, if cartilage calicifies during life, it has a much better chance of becoming fossil, like this fossil shark vertebra. In rare cases, other more delicate parts of sharks are preserved in the rock record.

Class:	Chondrichthyes
Family:	n/a
Habitat:	Marine
Distribution:	Global
Time scale:	Ordovician–Recent

Brachyrhizodus witchitaensis (teeth)

The rays are flattened bottom-dwelling members of the Chondrichthyes, the group of fishes which also includes shark, sawfishes and chimaeras. Rays are a very successful group – in terms of diversity, range and geologic longevity – due, in no small part, to their specialized dentition. The individual teeth of many rays are hexagonal or rhombic in outline and are geometrically packed together in the jaws with their flat surfaces contiguous to make two flat crushing surfaces. Rays swim along the sea floor and suck up hard-shelled prey like clams. The ligaments at the side of the jaws work just like a nutcracker's hinge and with the same results. The extremely durable teeth, like these from the Late Cretaceous Mount Laurel Formation of New Jersey, are common fossils in many marine deposits.

Class:	Chondrichthyes
Order:	Myliobatiformes
Habitat:	Marine
Distribution:	Eastern North America
Time scale:	Late Cretaceous

Myliobatoidei (barbs)

The stinger, or barb, of stingrays is a modified dermal denticle minute versions of which cover the entire body. The stinger, however, has modified its form considerably into a formidable weapon. The backwards-facing barbs make for a very painful removal, but the accompanying venom is a potent additional deterrent. The venom is not injected but instead is forced into the flesh of the target by the stinger breaking through its sheath of venom tissue and carrying these venom-laden cells into the entry wound. After losing one in a conflict, the ray quickly grows the next and readies it for use. These stingers, from Miocene Pungo River Marl, of North Carolina, show that it has been a successful device for millions of years.

Class:	Chondrichthyes
Order:	Myliobatiformes
Habitat:	Marine
Distribution:	Global
Time scale:	Cretaceous–Recent

Rhinobatos sp.

This now long-extinct creature was a member of a genus that still survives today – *Rhinobatos*, the guitarfish. Like its modern descendants, it had a skeleton of cartilage, so complete fossils of it are very rare. Far more common are fossils of its teeth, which were made from much harder material. *Rhinobatos* was like a cross between a shark and a ray. It had a flattened body and spent most of its time on the sea floor or the beds of rivers, like a ray. When it swam, however, it did so so by sweeping its tail from side to side like a shark. Rays swim by undulating the muscular fins along their sides. *Rhinobatos* almost certainly fed on shellfish and other small bottom-dwelling creatures. Its mouth was positioned on the flattened underside of its body and its teeth were blunt and rectangular.

Order:	Rhinobatiformes
Family:	Rhinobatidae
Habitat:	Seabed, estuaries and rivers
Distribution:	Global
Time scale:	Jurassic–Recent

Ischyrhiza mira (rostral denticles)

These 'teeth', more properly referred to as rostral denticles, come from the common Late Cretaceous sawfish *Ischyrhiza mira* and were found in the Mount Laurel Formation of New Jersey in the United States. They stuck out from the sides of a very elongated rostrum just like in modern sawfishes. They were also very likely to have been used in exactly the same way. But the similarity ends when the individual teeth are observed closely. These teeth have a flared root composed of two flattened attachments where the root connected to the snout. *Ischyrhiza* was a member of the sclerorhynchids, a group of sawfishes that evolved independently of the line that gave rise to the modern sawfishes, and represents a beautiful example of convergent evolution where similar lifestyles demand similar form.

Class:	Chondrichthyes
Family:	Sclerorhynchiae
Habitat:	Marine
Distribution:	North America
Time scale:	Cretaceous

Fish (coprolites)

Coprolites are fossil faeces and are abundant in many fossil deposits. It is usually very difficult to assign a coprolite to its maker since the animal is almost never found with the waste it expels. In some cases, the type of palaeoenvironment in which they are found can shed light on possible producers. Coprolites like these are found in marine sediments supporting a marine animal origin. The spiral exterior surface indicates that they were likely to be produced by fish with spiral valves in their intestines, such as sharks, rays, sawfishes, gars, lungfish, or coelacanths. All of these fish are found among the Late Cretaceous fossils of Big Brook, New Jersey, where these coprolites originate, but since sharks are by far the most common, they are the most probable producers of these distinct coprolites.

Order:	n/a
Family:	n/a
Habitat:	Marine
Distribution:	Global
Time scale:	Devonian–Pleistocene

Osteolepis sp.

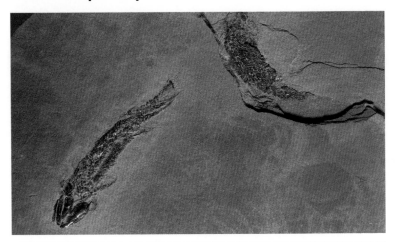

*O*steolepis was a lobe-finned fish, a member of the group that most scientists think gave rise to the amphibians and, through them, all other land vertebrates. *Osteolepis* was quite a small fish, growing to just 25cm (10in) long. Its name literally translates as 'bone scale' and refers to the heavy bony scales characteristic of the genus. It also had special bones in its fins, and these are the same bones that became the limb bones of terrestrial vertebrates. Over millions of years, they grew and changed in its descendants, slowly evolving into the bones of amphibian legs and feet. Although not our direct ancestor, *Osteolepis* was a close relative. Its powerful jaws were lined with dozens of tiny needle-sharp teeth, which it used to grab smaller fish and other prey.

Class:	Sarcopterygii
Order:	Coelacanthiformes
Habitat:	Fresh water
Distribution:	Europe
Time scale:	Devonian

Gyroptychius sp.

This fish lived during the Devonian Period. It was related to *Osteolepis* and the modern-day 'living fossil' known as the coelacanth. The type of rocks in which it is found show that *Gyroptychius* lived in fresh water rather than the sea. It probably inhabited rivers. *Gyroptychius* had a long streamlined body compared with most other fish of its time and was probably capable of sudden bursts of speed. Like most early fish, it was a meat eater. Its jaws were short but heavily muscled, giving it a powerful bite. *Gyroptychius* had small eyes and it probably found most of its food by smell. It grew to around 30cm (12in) long, making it similar in size to a modern-day trout. It also had the lobe fins typical of the group that gave rise to land vertebrates.

Class:	Sarcopterygii
Order:	Coelacanthiformes
Habitat:	Fresh water
Distribution:	Greenland & Europe
Time scale:	Devonian

Paranguilla tigrina sp.

*P*aranguilla tigrina was an eel that lived in the shallow seas covering what is now Italy during the Eocene. Other fossils found with it show it swam amongst tropical corals, suggesting that the region was much warmer then than it is today. Like eels today, *Paranguilla tigrina* had a long snake-like body and swam by 'wriggling' through the water, forming S-shapes with its body. *Paranguilla tigrina* was almost certainly an active hunter, catching and eating smaller fish. Most fossils of it show just the skeleton, but a few give the outline of its body, too. In a few rare fossils, there is even evidence of the patterns it had on its skin. It seems that *Paranguilla tigrina*, like the beautifully preserved specimen shown here, was covered with spots of colour like many moray eels today.

Order:	Anguilliformes
Family:	Anguillidae
Habitat:	Warm shallow seas
Distribution:	Europe
Time scale:	Eocene

Syngnathus sp.

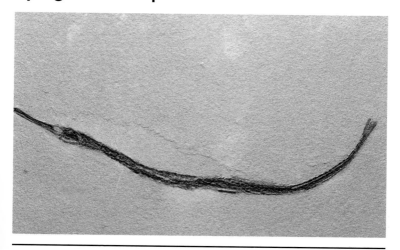

This little fish had very similar-looking relatives, which are still alive today. *Syngnathus* was a pipefish, a member of a group that is closely related to seahorses. It lived in shallow coastal waters in tropical and subtropical regions and was particularly common among seagrass beds, where its long slender body made it hard for predators to find it. *Syngnathus* ate tiny crustaceans and other creatures that lived in the water. Its mouth was very small, like that of a seahorse, and it fed by sucking in its prey rather than grabbing it between its jaws. Pipefish today live and feed in a similar way.

Order:	Gasterosteiformes
Family:	Syngnathidae
Habitat:	Coastal waters
Distribution:	Europe
Time scale:	Eocene–Recent

Eoplatax sp.

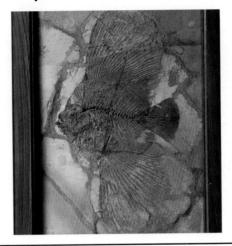

This bizarre-looking fish lived in tropical oceans around 45 million years ago. It was an early relative of today's batfish, which live in a similar way. *Eoplatax* probably spent much of its time around coral reefs, where it would have fed on small animals. Adults, such as this specimen, would have spent most of their time alone. Younger *Eoplatax* almost certainly travelled in shoals for protection, the mass of their moving bodies making it harder for predators to pick off one individual fish. *Eoplatax* skin rarely fossilizes, and even when it does it has no colour to it. Most modern batfish, however, are very brightly coloured and it seems likely that *Eoplatax* was very colourful too.

Class:	Actinopterygii
Order:	Perciformes
Habitat:	Tropical seas
Distribution:	Europe
Time scale:	Eocene

Notelops brama and *Rhacolepis buccalis*

The Santana Formation of Brazil is an extraordinarily rich deposit of Middle Cretaceous fossils whose species diversity, preservation, and abundance of fossils are legendary. Because of the minerals that preserve and encase the fossils, a technique called acid preparation can often be used to expose the delicate bony skeletons of the Santana. For the procedure to work, the rock must be susceptible to acid etching, but not the fossil. This acid-prepped specimen of *Notelops* exposes the remains of a *Rhacolepis* meal in its belly: the larger fish's vertebral column is on top and faces right and the head of the smaller is visible in the lower left. Rare specimens like these can be used to reconstruct trophic relationships, or the movement of nutrients within a community of organisms.

Order:	Crossognathiformes (*Notelops* and *Rhacolepis*)
Family:	Notelopidae (*Notelops*); Pachyrhizodontidae (*Rhacolepis*)
Habitat:	Marine
Distribution:	Ceará, Brazil
Time scale:	Middle Cretaceous

Semicossyphus pulcher (pharyngeal teeth)

When people hear the word 'teeth', they often picture structures lining the edge of the mouth used for biting or chewing. But for ichthyologists, the term teeth needs to be clarified because fish can have 'teeth' on the edges of the mouth, on the snout, the palate, and the tongue, in the throat, and even in the gills. Pictured here are batteries of pharyngeal, or throat, teeth from the sheepshead fish *Semicossyphus pulcher*, from the Miocene Santa Margarita Formation in California. They are semi-fused masses of tiny teeth that serve as a platform on which the fish crushes part of its hard-shelled diet. Specimens in the sandy deposit where these originate are always disarticulated, so the original correct identification of these depended on a broad knowledge of modern fish anatomy.

Order:	Perciformes
Family:	Labridae
Habitat:	Marine
Distribution:	Pacific USA
Time scale:	Eocene–Recent

Knightia sp.

*K*nightia was a fish related to herrings that could grow to a length of around 15cm (6in). Its fossils have been found in the freshwater sediments of North and South America. *Knightia's* body narrowed in shape from its somewhat short head towards the tail. The eye sockets were close together at the front – rather than at the sides – of the head, and the mouth was of average size. In the centre of its body, *Knightia* carried a triangular fin on its back, matched directly below it by the ventral fin on its underside. *Knightia* is an incredibly common fish in Wyoming's Eocene Green River Formation and some layers can be found with thousands of individuals of this one genus.

Class:	Actinopterygii
Order:	Clupeiformes
Habitat:	Lakes
Distribution:	USA
Time scale:	Eocene

Sphyraena bolcensis

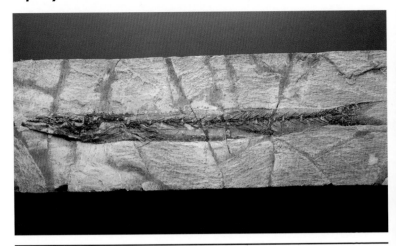

This fish shows the amazing ability of some creatures to survive almost unchanged over millions of years. *Sphyraena bolcensis* lived in the seas 45 million years ago, yet its skeleton looks almost identical to those of the barracuda which are its modern relatives. *Sphyraena bolcensis* was a fearsome predator that hunted other fish. Its body was torpedo-shaped and streamlined, enabling it to cut through the water at great speed when in pursuit of prey. Although we can never be certain, it seems likely that its body was silvery-blue in colour like those of its relatives today. This would have helped it to blend in with the background of open water, making it hard for its prey to spot it approaching until it was too late.

Order:	Perciformes
Family:	Sphyraenidae
Habitat:	Warm ocean waters
Distribution:	Europe
Time scale:	Middle Eocene

Cyclopoma spinosum

This fish lived in lakes and rivers around 45 million years ago. Despite its age, its skeleton looks very much like that of a modern perch, and that fish is indeed one of its living relatives. Not long after the non-avian dinosaurs died out, many of the fish families we know today had already evolved. Compare *Cyclopoma spinosum* with *Drepanaspis*, for example, and it is easy to see how much they had changed. *Cyclopoma spinosum* was almost certainly a predator, feeding on smaller fish and aquatic insects. As an adult, it probably spent most of its time on its own, lying in wait among the leaves and stems of water plants for prey. Young *Cyclopoma spinosum* would have been more likely to gather in shoals for protection from predators.

Class:	Actinopterygii
Order:	Perciformes
Habitat:	Fresh water
Distribution:	Europe
Time scale:	Eocene

Leptolepis sp.

At first sight, *Leptolepis* looks quite unremarkable. In fact, it looks like many fish that are alive today. What makes *Leptolepis* unique, however, is that it was one of the first of its kind. The reason it looks like so many modern fish is that it was one of the first bony fish or teleosts, the group that contains the vast majority of fish on Earth today. *Leptolepis* lived and behaved very much like a present-day herring. It fed on plankton, which it sieved from the sea with its gills, and lived in large shoals for protection from predators. Most of the creatures that fed on it belonged to groups that died out at the same time as the last of the non-avian dinosaurs. Its bones have been found preserved inside the skeletons of giant marine reptiles, including ichthyosaurs and plesiosaurs.

Order:	Pachycormiformes
Family:	Leptolepidae
Habitat:	Warm ocean waters
Distribution:	Europe & Asia
Time scale:	Jurassic

Vinctifger comptoni (in concretion)

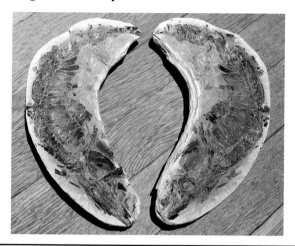

This fossil of the fish *Vinctifer* comes from the famous Santana Formation of Brazil. It is preserved in a concretion. These rock tombs are sometimes created when the chemicals of decay react with the surrounding sediments where a carcass is buried and cause cementation of the particles around the specimen. In many cases in the Santana Formation, this occurred rapidly enough to preserve soft tissues normally not preserved. And since concretions of this kind solidify early, they can protect a specimen from being flattened by the accumulated weight of later sediments, thus preserving the three-dimensional nature of the original. The concretions of the Santana conform so tightly to the contours of their contents that experts can often tell what species lies within before even splitting the rock.

Order:	Aspidorhynchiformes
Family:	Aspidorhynchidae
Habitat:	Marine
Distribution:	Ceará, Brazil
Time scale:	Middle Cretaceous

Mioplosus labracoides

*M*ioplosus was a powerful predatory fish that lived in freshwater habitats around 45 million years ago. The fact that its fossils are always found singly has led scientists to believe that it lived and hunted alone. *Mioplosus* probably hunted by stealth, lying in wait among water weeds for smaller fish to swim into range, then darting out suddenly to grab them before they had a chance to escape. It was related to the modern perch as were many other freshwater fish of the time, both in Europe, where it lived, and farther east in Asia and Australasia. Today, most members of the perch family live in the Northern Hemisphere. This specimen comes from the famous Eocene Green River Formation of Fossil Butte National Monument in Wyoming.

Class:	Actinopterygii
Order:	Perciformes
Habitat:	Lakes
Distribution:	USA
Time scale:	Eocene

Branchiosaurus sp.

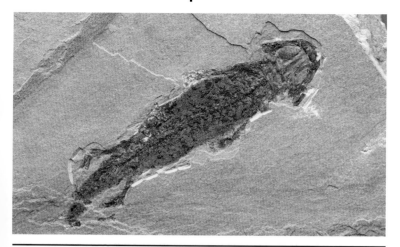

*B*rachiosaurus was an early amphibian. Its skeleton was not fully ossified: its spine, wrists and ankles were constructed mostly from cartilage. This intrigues early scientists who considered it to be a unique group of amphibians based on these shared features. But palaeontological thought grows and evolves over time just like the organisms it studies, and it now seems clear that *Branchiosaurus* has many of these characteristics because it is a larval form, probably of something like *Eryops*.

Class:	Amphibia
Order:	Temnospondyl
Habitat:	Fresh water
Distribution:	Europe
Time scale:	Late Carboniferous–Early Permian

Seymouria baylorensis

This small land animal used to be considered a 'missing link' between amphibians and other land-dwelling vertebrates. This is because *Seymouria* has an odd mix of characteristics that defined amphibians and reptiles. But these words have changed in meaning quite a bit in recent years. That said, *Seymouria* is still important to our understanding of the vertebrate transition to land. This stocky 'tetrapod' (vertebrates with four feet and legs) gets its name from Seymour, in Baylor County, Texas, USA, where the first examples were found.

Order:	Anthracosauria
Family:	Seymouriidae
Habitat:	Dry regions
Distribution:	Texas
Time scale:	Early Permian

Seymouriamorph (footprints)

Fossil footprints are notoriously hard to connect with their makers for a variety of reasons. One is that many closely related species can have very similar feet. Another is that we may know a fossil animal from its bones but the tracks show the foot clothed in flesh – a somewhat different form. Also, the consistency of the surface being trod has a great effect on the shape and detail of prints. But several things can help narrow down an identification. The age and environment of the tracks can be matched to known animals of similar time and place. Matching size is also helpful in reducing the number of possibilities. These tracks are assumed to be from seymouriamorph tetrapods – Permian terrestrial predators with aquatic larvae.

Order:	Anthracosauria
Family:	n/a
Habitat:	Terrestrial
Distribution:	Northern hemisphere
Time scale:	Carboniferous–Permian

Andrias scheuchzeri

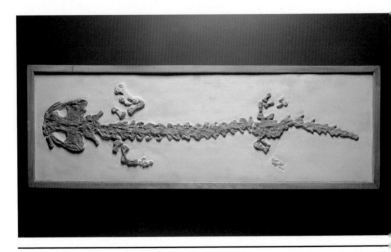

The hellbender salamanders of the eastern United States, and Asia's giant Chinese and Japanese salamanders are probably related to this now-extinct species. This ancient amphibian was first discovered in 1726 by Johann Jakob Scheuchzer (1672–1733), a Swiss physician. Scheuchzer believed that the fossil was the remains of a man who had drowned during the flood, described in the Christian Bible. He therefore named it *Homo diluvii testis*, meaning 'man, witness of the flood'. Later, the renowned French naturalist, Georges Cuvier (1769–1832), correctly identified the fossil as a giant salamander and renamed it *Andrias*, meaning 'image of man' and '*scheuchzeri*', in honour of its discoverer. *Andrias scheuchzeri* is currently on display in the famous Teylers Museum in Haarlem, in the Netherlands.

Order:	Urodela
Family:	Cryptobranchidae
Habitat:	Damp environments, such as mountain streams
Distribution:	Europe
Time scale:	Miocene

Rana pueyoi

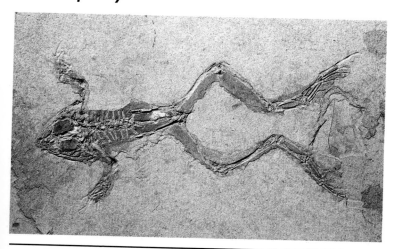

This extinct species of green frog made its home in the wet lands and swamps of North Eastern Spain, around 10 million years ago, during the Miocene Epoch. Recently these little amphibians have caused quite a stir in the scientific community. In most fossils, the only organic material that remains are bones, horns, teeth or shells – anything tough enough to survive the millennia. However, bone marrow has recently been extracted from the fossil remains of several species of frogs and salamanders, including *Rana pueyoi*. Bone marrow is found inside bone, and is responsible for helping to manufacture blood and fatty tissues. The importance of such 'soft tissue' finds is that it can tell us so much more about how an animal's body functioned, than just bone.

Order:	Anura
Family:	Ranidae
Habitat:	Fresh water
Distribution:	Most famous examples from Spain
Time scale:	Late Miocene

Frog

Complete frogs in the fossil record are very rare the majority of specimens are tiny isolated bones. But when found as whole skeletons, they tend to be neatly splayed out like this Early Miocene frog from Ar Rhyasha, Yemen. Most of the actual bones have disintegrated leaving a natural mold of the skeleton in the rock. Fossils like these can only provide morphological information as the internal structure of the bones is no longer available. The dark, weathered, original rock surface can clearly be seen. This contrasts with the lighter, freshly-exposed rock removed during preparation of the fossil. Also visible are dark registration marks drawn by the collector on the individual rock chunks. This was done to facilitate reassembly of the parts in the lab when it was time to clean the fossil.

Class:	Amphibia
Order:	Anura
Habitat:	Freshwater & terrestrial
Distribution:	Global
Time scale:	Triassic–Recent

Labidosaurus hamatus

*L*abidosaurus hamatus was one of the world's first reptiles. It lived 300 million years ago, 50 million years before the first dinosaurs appeared. Unlike most modern reptiles, *Labidosaurus hamatus* had a wide flattened head. In this respect, it was more like an early tetrapod (the group from which it and other reptiles evolved). *Labidosaurus hamatus* was also one of the world's first true land vertebrates. It had dry scaly skin and laid eggs with tough shells. This meant that, unlike the earliest tetrapods, it was able to spend much of its time away from water. This enabled it and its relatives to move out of the swamps and damp forests into drier habitats where there was no competition for food. *Labidosaurus hamatus* was about the size of a small dog and fed mainly on insects and other invertebrates.

Order:	Captorhinida
Family:	Captorhinidae
Habitat:	Forests, plains & possibly deserts
Distribution:	North America
Time scale:	Late Permian

Trionychid

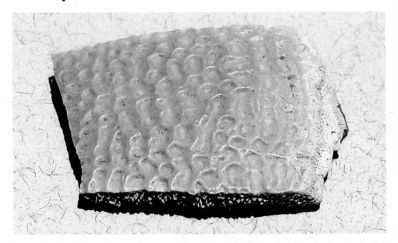

The earliest known turtles date back to the Triassic. But since they are by that time true turtles, they shed little light on the ancestry of the group. Turtles have been, and are, an enormously successful group, and the fact that the basic turtle form has changed little in over 230 million years attests to this. One very successful turtle lineage is the trionychids, or soft-shelled turtles. Like all turtles, the carapace (the shell of their back) is composed of the fused ribs of their skeleton. Trionychids are considered 'soft' because of the flexible margin of their shell and the lack of horny scutes on the carapace. The pitted carapace bones of trionychids, like this Late Cretaceous specimen from the Hell Creek Formation of South Dakota, are common fossils from fresh water deposits from the Cretaceous forward.

Order:	Testudines
Family:	Trionychidae
Habitat:	Fresh water
Distribution:	Global
Time scale:	Cretaceous–Recent

Stylemys nebrascensis

*S*tylemys nebrascensis was a tortoise that lived in North America approximately 30 million years ago. It is one of the most frequently found fossils in North America from rocks laid down in the Oligocene Period, suggesting that it was very common when it was alive. *Stylemys nebrascensis* was a herbivore, feeding on plants that grew close to the ground. Like modern tortoises, it was probably quite long-lived, with individuals surviving for 100 years or more. Some *Stylemys nebrascensis* grew very large, with shells of up to 1m (3¼ft) long. In the wild, it would have had few predators and little to fear from any animal that tried to attack it. Like tortoises today, it could pull its head and legs inside its tough shell for protection.

Order:	Testudines
Family:	Testudinidae
Habitat:	Forests & plains
Distribution:	North America
Time scale:	Oligocene

Geochelone osborniana (tail armour)

Turtles and tortoises are famous for their enclosing shells, composed of the fused ribs of the animal's torso and other bones. This armour is unparalleled in the animal world. But it is not all the armour the group has worn. Tortoise limbs can be heavily scaled, and these scales commonly house bony nuggets called osteoderms. This *Geochelone* specimen found in the Miocene Pawnee Creek Formation near the state line between Nebraska and Colorado shows further osteoderm armouring that covered even *Geochelone's* tail. A good knowledge of various elements of diverse bony skeletons is necessary in vertebrate palaeontology since most skeletons disarticulate after death: these osteoderms, especially if isolated, would be hard to identify without prior recognition of them as part of the armour of a tortoise.

Order:	Testudines
Family:	Testudinoidea
Habitat:	n/a
Distribution:	Global
Time scale:	Eocene–Holocene

Macrocnemus bassani

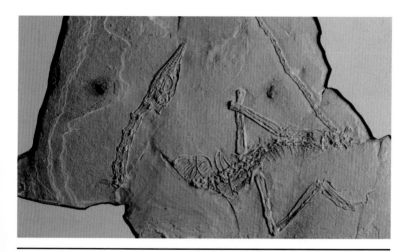

Straddling the border of southern Switzerland and northern Italy is a rich vertebrate fossil site called Monte San Giorgio. It preserves in great detail marine vertebrates of the Triassic that lived in an environment very similar to the modern Bahamas. Occasionally, during storm events, terrestrial species would also be swept into the basin and drowned, like *Macrocnemus*, a small terrestrial relative of lizards that is commonly found in these sediments. Vertebrate fossils of this deposit are almost always found complete, and *Macrocnemus* is no exception. Some specimens have even been found with the delicate remains of their scaly skin. The proportions of *Macrocnemus*' limbs suggest that, at least some of the time, it was able to run on just its hind limbs, but most of the time, it is likely to have moved around on all fours.

Order:	Protosauria
Family:	Prolacertidae
Habitat:	Coastal terrestrial
Distribution:	Europe
Time scale:	Triassic

Mesosaurus brasiliensis

These small reptiles lived in lakes and streams during the Permian Period. With its streamlined body, webbed feet and finned tail, *Mesosaurus* was a swift swimmer, able to use its undulating tail to propel itself forwards through the water. However, its thickened ribs would have probably made sideways motion – and therefore sudden acrobatic turns – during the pursuit of prey difficult to achieve. *Mesosaurus brasiliensis* fossils have played a vital role in proving that our Earth's land masses constantly move. Fossils have been found in South America and South Africa. As this aquatic reptile lived in freshwater, and would be unable to cross the Atlantic Ocean, it proved that the two continents must once have been joined in one 'super continent' (called Pangaea).

Order:	Mesosauria
Family:	Mesosauridae
Habitat:	Freshwater
Distribution:	South America and South Africa
Time scale:	Permian

Placodus sp. (teeth)

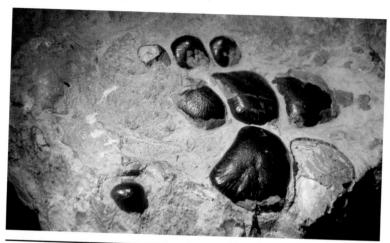

This rock from the Triassic of Windlack, Germany, displays an odd pattern of teeth. They are the large palatal crushing teeth of the typical, and namesake, placodont *Placodus*. Placodonts only lived during the Triassic and, so far, have been found in Europe and northernmost Africa. In the short time they existed, they managed to evolve a variety of body forms ranging from iguana-like to turtle-like with heavy body armour. They were amphibious marine animals and presumably used their heavy teeth to crush the hard-shelled invertebrates they ate along the sea shore. Given the fact that the teeth exposed on the surface of this rock are in the pattern found in the animal's palate, it is likely that further preparation would produce a well-preserved skull.

Order:	Placodontia
Family:	Placodontidae
Habitat:	Coastal marine
Distribution:	Europe and northern Africa
Time scale:	Middle–Late Triassic

Ichthyosaur

Ichthyosaur is Greek for 'fish lizard' and, as its name suggests, this marine reptile evolved from a reptile that returned to the sea. This is especially clear when you compare the fossils of advanced ichthyosaurs with their ancient predecessors, which looked like land reptiles with fins. Despite having to surface to breathe air, ichthyosaurs were fearsome predators. Their streamlined shape made them fast and agile, while large eyes made it possible for them to hunt well in the gloom of the ocean depths. These natural advantages made these giant hunters which sometimes grew up to 18m (60ft) long, one of the top marine hunters of the Jurassic Period. The first complete ichthyosaur skeleton was found by 12-year-old Mary Anning (1799–1847) in 1811.

Order:	Ichthyopterygia
Family:	n/a
Habitat:	Marine
Distribution:	Global
Time scale:	Mesozoic

Stenopterygius quadriscissus (birth)

This extraordinary fossil shows the ichthyosaur *Stenopterygius quadriscissus* in the act of giving birth. The mother's head is out of the frame to the left and the tail extends off to the right. Underneath her large spine are the vertebral columns of some of her unborn and the complete skull of a newborn can be seen exiting her body on the lower right. It cannot actually be said whether this parent died while giving birth or if the foetuses were expelled after death, but it does prove that these animals had live birth, forsaking the egg-laying habits of their terrestrial ancestors. This fits well with their very advanced adaptation to a marine existence.

Order:	Ichthyopterygia
Family:	Stenopterygiidae
Habitat:	Marine
Distribution:	Holzmaden, Germany
Time scale:	Early Jurassic

Askeptosaurus italicus

It is easy to see from the shape of its skeleton that *Askeptosaurus italicus* was an aquatic reptile. With an eel-like body that was 2m (6½ft) long, a muscular tail and broad webbed feet, *Askeptosaurus* would have been a fast and formidable hunter. Large eyes, surrounded by a circle of bone possibly to protect them from water pressure, would have enabled this sleek predator to dive to considerable depths in pursuit of its prey. Typical food would have been fish, which it would have torn apart with rows of sharp teeth. Fossils of *Askeptosaurus italicus* have, most famously, been found in Monte San Giorgio, Switzerland. This region is so rich in fossil finds that it has been awarded the status of 'World Heritage Site' by the United Nations Educational, Scientific and Cultural Organisation (UNESCO).

Order:	Thalattosauria
Family:	Askeptosauridae
Habitat:	Marine
Distribution:	Switzerland
Time scale:	Triassic

Ceresiosaurus sp.

Anyone who knows the legend of the Loch Ness Monster will be familiar with the design of these long-necked aquatic reptiles. Yet, while *Ceresiosaurus* was a skilled hunter, it was no prehistoric monster. In fact, at just 4m (13ft) long, it would be dwarfed by the creature that inspired the Nessie myth – a plesiosaur, which could reach 14m (46ft) long. However, of all the nothosaurs, *Ceresiosaurus* is perhaps the most plesiosaur-like, with four fully developed flippers and a small compact head. While most nothosaurs moved through water using a snakelike 'undulating' movement, *Ceresiosaurus* was a much more agile and accomplished swimmer, moving a little like a penguin. It was wiped out 150 million years ago during the Triassic-Jurassic extinction event, which killed 20 per cent of all marine animals.

Order:	Nothosauria
Family:	Nothosauridae
Habitat:	Marine
Distribution:	Europe
Time scale:	Triassic

Mosasaur (tooth)

This fossil shows the tooth of an ancient sea reptile known as a mosasaur. Mosasaurs lived during the Cretaceous Period (from around 144–65 million years ago). They were the last of the giant marine reptiles to appear, taking to the oceans long after the first ichthyosaurs and plesiosaurs. Mosasaurs were meat-eaters and some of them grew very large. *Mosasaurus*, for instance, was over 15m (50ft) long – big enough to hunt down and kill most other sea reptiles. Mosasaurs, like plesiosaurs, died out in the mass extinction event 65 million years ago which also killed off the non-avian dinosaurs. Before they disappeared however, they had colonized coastal waters across most of the earth and had become an extremely successful group of ocean predators.

Order:	Squamata
Family :	Mosasauridae
Habitat:	Coastal Marine
Distribution:	Global
Time scale:	Cretaceous

Crocodylian (osteoderm)

Crocodilians have bony armour that covers their entire back. Under a thin covering of skin are many bones called osteoderms, which are the units of this covering and have an ornamented surface common to bones that directly underlie skin without intervening muscles or fat. They are very common as fossils because crocodiles have a long history, were very widely distributed, and each animal had numerous osteoderms. Only relatively recently has it become known that in modern crocodilians these bones in the skin serve an important role in locomotion and are tightly linked to the workings of the muscles and ligaments of the back. Top predators have little use for such heavy armament, but it is possible that the earliest crocs used them as armour and their locomotory function evolved later.

Class:	Reptilia
Order:	Crocodylia
Habitat:	Aquatic
Distribution:	Global
Time scale:	Late Cretaceous–Recent

271

Crocodylian (jaw in concretion)

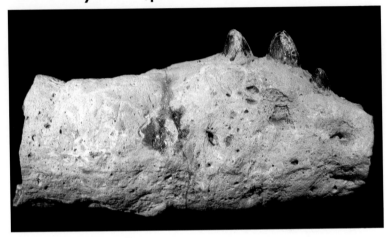

As fossils weather out of their host rock, they can often be encrusted with a patina of rock that is harder than the surrounding matrix. This can be in the form of a concretion, which is a rock formed by the cementing of enclosing sediments often due to the presence of decay byproducts. This concretion exposes a line of teeth, which indicate that a jaw lies within – the jaw of a small crocodile. The fossil was found in the Palaeocene rocks of Los Barreales, New Mexico, and was probably lying on the surface as seen here (except for the glued crack across the middle). When concretions like this are found exposing parts of a fossil skeleton, nearby concretions may contain more of the same weathered skeleton and should be checked for enclosed remains.

Family:	Reptilia
Order:	Crocodylia
Habitat:	Aquatic
Distribution:	Global
Time scale:	Cretaceous–Recent (true crocodiles)

Tapejara wellnhoferi

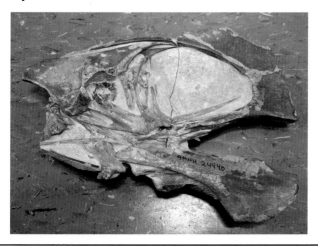

Pterosaurs were a group of flying vertebrates very closely related to dinosaurs, of which the birds are members. But pterosaurs evolved a unique wing design – a solution to flying very different to that developed by birds. The fourth finger of the hand was greatly elongated and supported a flight membrane. Birds' wings are made from a reduced and partially fused hand skeleton internally, and feather extensions externally. The tapejarids were one of the most bizarre groups of pterosaurs and developed a variety of odd crests of unknown function. This beautiful *Tapejara* skull shows its deep face and some of the expanded front end of the skull and lower jaws, creating two opposing crests somewhat resembling a hatchet. Also preserved is the delicate circle of tiny eye bones called the sclerotic ring.

Order:	Pterosauria
Family:	Tapejaridae
Habitat:	Coastal
Distribution:	Brazil
Time scale:	Early Cretaceous

Diplodocid and Allosaur (toothmarks)

Trace fossils, such as the tooth marks seen here as scratches and gouges on a sauropod bone from the Late Jurassic Bone Cabin Quarry in Wyoming, preserve the activity of ancient organisms. It is clear from the teeth of the large bipedal dinosaurs of the time that they were meat eaters, but this kind of fossil lends dramatic credence to this observation. It is tempting to picture a huge predatory theropod killing and tearing into a giant, but overpowered, plant eater, but a scavenging theropod would probably leave exactly the same kind of marks. All that can be said with reasonable confidence is that a theropod was feeding on a sauropod. However, it is not out of the realm of possibility that a theropod could have had other reasons for gnawing on sauropod bones.

Order:	Saurischia
Family:	Diplodocidae (bitten bone); Allosauridae (bone biter)
Habitat:	n/a
Distribution:	Wyoming (this specimen)
Time scale:	Late Jurassic (this specimen)

Psittacosaurus sp.

Although not as famous as other 'ceratopsian' (horned) dinosaurs like *Triceratops*, these primitive vegetarians are well known to palaeontologists due to the large number of fossil skeletons that have been found. So far, eight separate species have been discovered, including complete nests, which suggest that females may have taken it in turns to care for the herd's young. Although size varies between species, the largest examples of *Psittacosaurus* grew to around 2m (6½ft) long, 1.2m (4ft) tall and weighed in at 80kg (175lbs). Dating from around 130–100 million years ago, *Psittacosaurus* is easily identified by its beak-like jaws, which were probably used to tear leaves. It is this beak that gives the genus its name – 'parrot lizard'.

Order:	Ornithischia
Family:	Psittacosauridae
Habitat:	Terrestrial
Distribution:	China, Mongolia & Thailand
Time scale:	Early Cretaceous

Edmontosaurus sp.

If these rocks were found on their own, they could not be considered fossils. But the fact that they were found associated with the fossils skeleton of a dinosaur makes them probable gastroliths – stones ingested by an animal – and thus a type of trace fossil. There are many reasons an animal, in this case an herbivorous hadrosaur, might do this. It might use them to grind its food in a gizzard. They might be eaten for their mineral content. They might have been eaten by accident along with the animal's normal food. No one knows for sure. The same is true for the animals today that do it. Sometimes the rock-types eaten originate far from where the skeleton was found and suggest a migratory animal.

Order:	Ornithischia
Family:	Hadrosauridae
Habitat:	Terrestrial
Distribution:	Global
Time scale:	Cretaceous

Non-avian Dinosaur (Egg)

Fossilized eggs belonging to almost all of the major non-avian dinosaur groups have been discovered, the first to be found belonging to the giant, long-necked *Hypselosaurus*. The smallest fossilized dinosaur eggs are those of the tiny mouse-like *Mussaurus,* which are just over 2.5cm (1in) across. The importance of fossilized dinosaur eggs lies in what they can potentially tell us about non-avian dinosaur life cycles. Non-avian dinosaurs are closely related to both birds, their descendants, and, less so, crocodiles, yet it is still unclear if they laid their eggs one at a time, like birds, or all at once, like crocodiles. However, eggs containing unhatched embryos have already given palaeontologists important clues about non-avian dinosaur anatomy and development.

Order:	n/a
Family:	n/a
Habitat:	n/a
Distribution:	Global
Time scale:	Late Triassic–Recent

Titanosaur (eggshell fragments)

These eggshell fragments are from titanosaur eggs found in the Late Cretaceous Anacleto Formation of Neuquén Province, Patagonia. Titanosaurs are a lineage of sauropod dinosaurs that included some of the largest animals ever to walk on land. Some of them surpassed 30m (100ft) in length and weighed as much as 80 tons. However, titanosaur eggs are surprisingly small: the bits shown here come from eggs 15cm (6in) in diameter. The unexpectedly small size of these eggs is due to the fact that as eggs increase in size, the thickness of the shell must increase accordingly to resist its own weight crushing it. But beyond a certain thickness, the embryo within will not be able to breath because gasses can no longer easily pass through the shell.

Order:	Saurischia
Family:	Titanosauridae
Habitat:	Terrestrial
Distribution:	Global
Time scale:	Cretaceous

Titanosaur (embryo skin)

A uca Mahuevo, in Patagonia's Anacleto Formation, preserves a most
extraordinary fossil site. In the Late Cretaceous, a flood plain, dotted with the
egg-filled nests of titanosaurid dinosaurs, was inundated by a muddy flood. What was
very unfortunate for these dinosaurs was exceptionally lucky for palaeontologists.
Although fossil dinosaur eggs are relatively common, they are rarely preserved right
before hatching. Even more rare is that they are mineralized in a way that preserves
the skeletal contents. But rarest of all is the preservation of the skin of the embryonic
dinosaurs. A small patch can be seen at the top centre of this partial egg: the dark
curve on the bottom is the edge of the preserved shell. Auca Mahuevo is the only
place on Earth where the fossils of embryonic dinosaur skin have ever been found.

Order:	Saurischia
Family:	Titanosauridae
Habitat:	Terrestrial
Distribution:	Global
Time scale:	Cretaceous

Theropod (footprint)

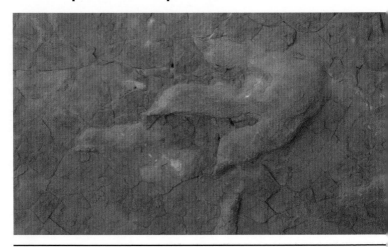

Not all fossils are of shells or bones. Some show where an organism has been or what it has done without preserving any remains of that animal itself at all. Fossils like these are known as trace fossils. Dinosaurs left many trace fossils in the form of footprints. These tell us things we may not have otherwise known about the creatures that left them, such as how flesh was distributed around the bones of their feet. Occasionally footprints are fossilized together, forming tracks. Knowing the lengths of the legs of the animals that made them, these tracks give us an idea of how fast the creatures moved. Sometimes dinosaur tracks tell us even more. Where the footprints of a theropod, like the meat eater's track shown here, follow those of another dinosaur, we can guess that they were left by a predator chasing after prey.

Class:	Sauropsida
Order:	Saurischian (this specimen)
Habitat:	All land habitats
Distribution:	Global
Time scale:	Triassic–Recent

Archaeopteryx lithographica

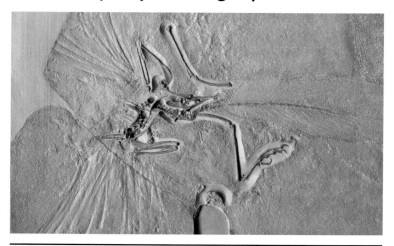

It is now generally agreed that birds evolved from a group of carnivorous dinosaurs, some time during the late Late Jurassic period, about 150 million years ago. The name, *Archaeopteryx lithographica*, means 'ancient wing from a printing stone'. This refers to a type of limestone, which was used in printing, in which the earliest fossilized feather was found. Although *Archaeopteryx* had feathers, it also had many more 'reptilian' features, such as teeth, claws, and a bony tail. From what is known about modern birds, it seems likely that this magpie-sized 'proto-bird' was a glider rather than a true flier. Ten specimens of *Archaeopteryx* have been found, and debate rages as to whether or not they belong the same species.

Order:	Saurischia
Family:	Archaeopterygidae
Habitat:	Coastal regions
Distribution:	Germany
Time scale:	Late Jurassic

Diatryma gigantea (beak part of skull

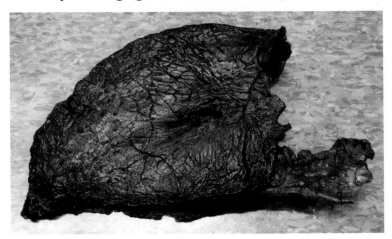

Diatryma was a giant flightless bird about 2m (6½ft) tall; this beak is 13cm (5in) in longest dimension. Usually, when people think of dinosaurs, they imagine giant animals that became extinct 65 million years ago. But now that we know birds are dinosaurs, *Diatryma* can be seen as a giant dinosaur that was around well after 65 million years ago. In fact, the largest known bird was a recently discovered Miocene giant ground-dwelling predator that stood over 3m (10ft) tall, which puts a truly giant predatory dinosaur on Earth 15 million years ago. Called 'terror birds', the group persisted until about two million years ago. *Diatryma* is traditionally thought of as a carnivore, but it lacks the hooked beak of predatory birds and is most closely related to ducks.

Order:	Saurischia
Family:	Diatrymatidae
Habitat:	Terrestrial
Distribution:	Europe & North America
Time scale:	Eocene

Aves (feather)

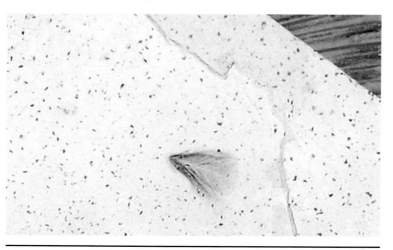

This tiny Eocene feather comes from the Green River Formation of Douglass Pass, Colorado. Because of the fine grain and layered nature of the rocks at Douglass Pass, a year's worth of freezing and thawing can expose thousands of beautiful fossils that merely need to be collected as is. Feathers used to be a clear indication of birds, but recent discoveries of the dinosaurian relatives of the first bird show that feathers existed in many dinosaurs we would not call birds. Therefore, if one were to find a feather from the Late Jurassic or Cretaceous, it could not be said with certainty that it was from a bird. This Eocene feather, on the other hand, is doubtless from a bird since it comes from a time after the extinction of the last non-avian dinosaurs.

Order:	n/a
Family:	n/a
Habitat:	n/a
Distribution:	Global
Time scale:	Late Jurassic–Recent (feathers)

Edaphosaurus sp. (vertebra)

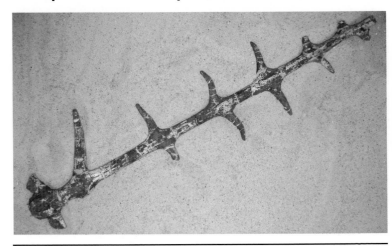

Pelycosaurs were the earliest known synapsids – the group of tetrapods today represented by the mammals. The group is likely to be paraphyletic (artificially composed of several independent lineages instead of representing a single lineage with a hypothetical ancestor and all of its descendants). *Edaphosaurus* was a fin-backed herbivorous pelycosaur reminiscent of the contemporary carnivore *Dimetrodon,* which also sported a huge sail on its back. These pelycosaurs lived in Texas, USA, and it is likely that *Dimetrodon* preyed on *Edaphosaurus.* A significant difference in the two animals, aside from diet, was the sail structure: *Dimetrodon's* was supported by simple rods of bone rising from the back vertebrae; *Edaphosaurus'* rods, visible here, had lateral projections of unknown function.

Order:	Pelycosauria
Family:	Edaphosauridae
Habitat:	Terrestrial, possibly near swamps
Distribution:	Texas, USA
Time scale:	Permian

Brontotherium sp.

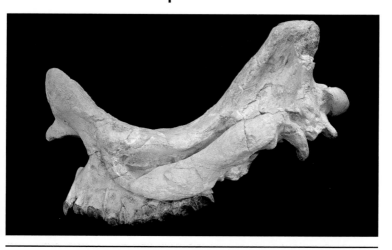

Sioux tribes, who found the bones of these great mammals exposed after rain storms, believed that *Brontotherium* were responsible for making thunder by running across the clouds. They therefore gave them the name 'thunder beasts', which is what *Brontotherium* means in Greek. This impressive creature stood at around 2.5m (8ft) tall, with a broad body, four columnar legs and thick skin, like a rhinoceros. Its most prominent feature was a huge Y-shaped horn. This was larger in males, suggesting that it was used during combats for mates. *Brontotherium* is an odd-toed ungulate, with four hoofed toes on its front legs and three on its back legs. The fossil shown here is part of a *Brontotherium's* skull.

Order:	Perissodactyla
Family:	Brontotheriidae
Habitat:	Plains
Distribution:	North America
Time scale:	Early Oligocene

Brontops sp. (fracture)

Palaeopathology is the study of injury and disease in the fossil record. In a large sample of a single genus or species, palaeontologists may be able to determine the commonness and distribution of certain pathologies from the fossils. But other interesting details can be hinted at by fossil pathologies as well. This mounted skeleton of *Brontops robustus* shows a bad but partially healed fracture to one of its right ribs. Healing can prove that an injury was non-fatal, but injuries of this kind may also shed light on behaviour. Breaks on the ribs of animals that fight by locking horns or other cranial projections show that side-butting was also a likely behaviour. *Brontops* males probably used their horns in a similar way when battling for females.

Order:	Perissodactyla
Family:	Brontotheriidae
Habitat:	Terrestrial
Distribution:	Western USA
Time scale:	Eocene

Miohippus sp.

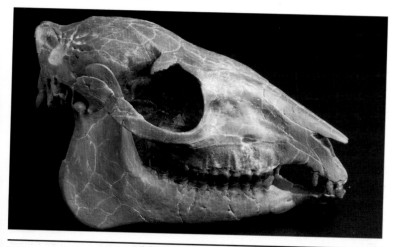

These early ancestors of the modern-day horse were first discovered by American palaeontologist Othniel Charles Marsh (1831–1888) in 1874. At the time, Marsh believed that the fossils he had found came from rock formations laid down during the Miocene Epoch. He therefore gave them the name 'Mio' from Miocene and 'hippus' meaning 'horse'. It is now known that *Miohippus* is actually much older – dating from the Oligocene Epoch (34 million years ago). Two species of these agile pre-horses have so far been discovered – one made its home on the plains, the other preferred woodlands. Compared to modern horses, which weigh in at around 500kg (1100lbs), *Miohippus* was tiny, tipping the scales at just 55kg (120lbs).

Order:	Perissodactyla
Family:	Equidae
Habitat:	Forests & prairies, depending on species
Distribution:	North America
Time scale:	Mid-Oligocene

Arsinoitherium sp.

The most distinctive features of these stocky mammals are the pair of long hollow horns on their snouts. Some think that these may have acted like 'loud hailers', magnifying the male's call to attract females during the mating season. Bulky Arsinoitherium reached 1.8m (6ft) at the shoulder and was a herbivore which spent much of its time grazing. As it couldn't straighten its legs completely, it has been suggested that *Arsinoitherium* was 'semi-aquatic' – wading along the water's edge as it fed. In common with the rhinoceros, which it resembled, *Arsinoitherium* may have lived together in small family groups for protection from predators. The name *Arsinoitherium* comes from the Egyptian Queen Arsinoe III (246–204BC), whose ancient palace was close to the site of the first fossil find.

Class:	Mammalia
Family:	Arsinoitheriidae
Habitat:	Terrestrial
Distribution:	North Africa
Time scale:	Eocene–Early Oligocene

Glyptodon clavipes (osteoderm)

R elatives of armadillos, glyptodonts were huge, heavily armoured herbivores that populated the Americas for most of the Cenozoic. Their backs and sides were completely encased in a dome of bone made up of a mosaic of small fused bones called osteoderms, like the Pleistocene specimen from South America pictured here. Their tails were also sheathed in bone and, in some species, the end of the tail was a mace of heavy spikes. Even the heads of some species were protected on top by a cap of armour called a casque. This heavy armour surely slowed them down, but also made them nearly impenetrable to most predators. Originating in South America, glyptodonts made it to North America after the land bridge of Central America was formed some time in the Pliocene.

Order:	Cingulata
Family:	Glyptodontidae
Habitat:	Terrestrial
Distribution:	North & South America
Time scale:	Eocene–Pleistocene

Sloth (tracks)

Fossilized tracks tell us a lot about the creatures that made them. Their depth gives us an idea of the animal's weight and the distance between the footprints, its speed and the length of its stride. Tracks also tell us how many legs an animal had. Shown here is a small portion of an area of trackways in Patagonia. These may look like modern tracks in the mud of a beach, but they actually preserve tracks of giant sloths and other extinct animals of the Pleistocene. Since the tracks are on a present beach, they are easily buried and scientists studying them must frequently unbury them.

Order:	Xenarthra
Family:	Megatheriidae
Habitat:	Terrestrial
Distribution:	Global
Time scale:	Miocene–Recent

Megatherium sp.

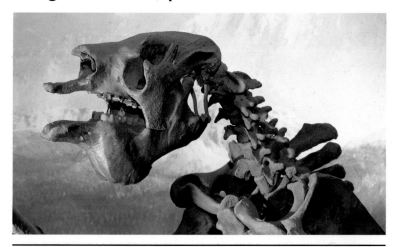

***M**egatherium* was a giant sloth that lived in the Americas. Extinct for the last 11,000 years, it possessed a huge high skull and deep jaws containing peg-like teeth. Despite its fearsome appearance, *Megatherium* was a herbivore, eating leaves, shoots and fruits. Known as the giant ground sloth, *Megatherium* (meaning 'Great Beast') could grow to a length of 6m (20ft). Thick fur gave protection against the deep freeze of the Ice Ages and all four feet carried large claws, curved upwards on the back pair and backwards on the front pair. Although primarily a quadruped, *Megatherium* could stand on its hind feet, using its thick tail as a balance, and reach for foliage high up in the trees.

Order:	Xenarthra
Family:	Megatheriidae
Habitat:	Terrestrial
Distribution:	Global
Time scale:	Miocene–Recent

Hyaenodon sp.

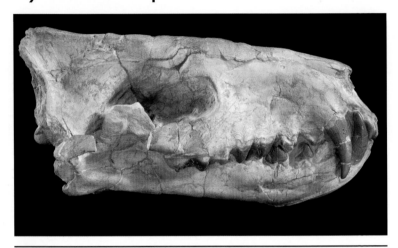

As can be seen from this fossilized skull, *Hyaenodon* was a hunter and a carnivore. The largest member of the genus, *Hyaenodon gigas* measured up to 1.4m (4½ft) tall, and was probably a solitary predator. Smaller species most likely hunted in packs like modern wolves. *Hyaenodon* means 'hyena toothed'. This is a reference to its large and powerful teeth which, like those of a hyena, are designed to crush bones. *Hyaenodon* belonged to a group of mammals known as 'creodonts', which died out around 21 million years ago. It is believed that creodonts were driven to extinction by the arrival of a group of more successful hunters that belonged to the group Carnivora. These were the ancestors of today's true carnivores.

Order:	Creodonta
Family:	Hyaenodontidae
Habitat:	Forests & plains
Distribution:	North America, Europe, Africa, Asia
Time scale:	Late Eocene–Miocene

Hoplophoneus sp.

This extinct North American predatory mammal is often mistakenly listed as the ancestor of *Smilodon* – popularly known as the sabre-toothed tiger. However, it is now believed that *Hoplophoneus* developed before *Smilodon*, during the Oligocene Period (35 million years ago). Additionally, and, despite its feline-like appearance, *Hoplophoneus* belonged not to the cat family (Felidae) but to a group of predatory carnivores called nimravids. Five species of *Hoplophoneus* have so far been discovered, with examples ranging in size from that of an average bob cat to a fully grown jaguar. Its most prominent feature – those curved, sickle-like teeth – serve an unknown function, but they were probably too fragile to be weapons. When not in use, the teeth rested in a specially designed bony ridge in the jaw.

Order:	Carnivora
Family:	Nimravidae
Habitat:	Prairies & grasslands
Distribution:	North America
Time scale:	Eocene–Oligocene

Ursus spelaeus (tooth)

S tanding on its hind legs, this gigantic bear measured an average of 3m (10ft) tall – that's around 30 per cent larger than the modern-day brown bear. These huge omnivores get their name from the fact that they appear to have lived in caves all year round, unlike today's bears who only use them for shelter during the winter, when they hibernate. Cave bear remains, like this tooth, have been found in abundance at some sites, leading to the suggestion that that early humans may have used bear bones in religious rites. Some Neanderthal caves have even been found with bear skulls placed on alters, suggesting that they were revered by the very people who may have ultimately driven them to extinction.

Order:	Carnivora
Family:	Ursidae
Habitat:	From mountain regions to lowland forests
Distribution:	From Britain to Central Europe
Time scale:	Pleistocene

Archaeotherium sp.(tooth)

*A*rchaeotherium was an early relative of today's hoofed mammals. Its name means 'ancient beast'. *Archaeotherium* lived around 35 million years ago. In many ways, it was similar to the modern-day warthog, although it was slightly larger – closer in size to a cow. Like modern pigs, *Archaeotherium* fed mainly on plant material, but it also scavenged carrion. It had powerful neck muscles that attached to the shoulder blades to form a slight hump behind its head. These would have enabled it to snuffle around with its snout and dig up plant tubers and other nutritious roots. Some *Archaeotherium* skulls have been found with deep gashes in them. These were almost certainly caused during fights with others of their own kind, probably between males as they battled for the right to mate with females.

Suborder:	Suina
Family:	Entelodontidae
Habitat:	Forests
Distribution:	North America
Time scale:	Eocene–miocene

Protoceras sp.

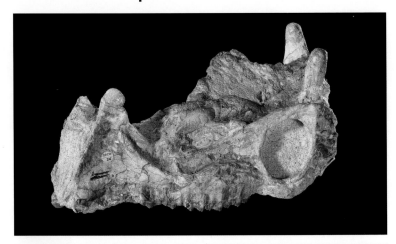

Protoceras was an early hoofed mammal related to modern-day antelopes and deer. It lived around 20 million years ago and had four hoofed toes on each foot. *Protoceras* was quite small compared to most hoofed mammals alive today, growing to just 1m (3¼ft) long. Its name means 'first horns'. In life, these were probably covered with skin, like the horns of a giraffe. Males had three pairs of horns, one pair behind the eyes, one large, flattened pair near the front of the snout and a third pair of small horns in between. Females had just one pair of horns, those behind the eyes, which were smaller than the same horns on males.

Order:	Artiodactyla
Family:	Protoceratidae
Habitat:	Plains & woodlands
Distribution:	North America
Time scale:	Late Oligocene–Early Miocene

Bison sp. (horn cores)

*B*ison have been a feature of European and North American grasslands for over four million years. Immense herds containing millions of individuals once roamed freely between the North American Appalachian Mountains and the Rockies. However, in the nineteenth century, many millions were systematically slaughtered, with the result that there were fewer than 1000 left in the wild by 1889. Today they are protected throughout the United States where they are commonly known as 'Buffalo'. US Bison are traditionally large – up to 3.6m (12ft) tall – with coarse brown-black hair, a distinctive humped back and a pair of heavy curved horns. Their European relatives, which can be found on farms throughout Eastern Europe, Italy and Asia Minor, are smaller with shorter hair.

Order:	Artiodactyla
Family:	Bovidae
Habitat:	Temperate grasslands
Distribution:	Europe, North America & Asia Minor
Time scale:	Pliocene Epoch–Recent

Megaceros giganteus

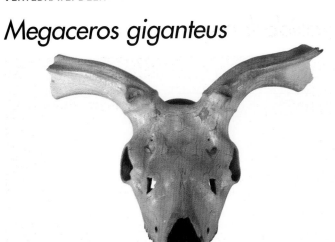

Megaceros giganteus is sometimes called the 'irish elk'. This is despite the fact that it is neither an elk nor exclusively Irish, although skeletons have been found in abundance in Irish bogs and marl pits. A more accurate description for this huge beast is the 'giant deer', due to its enormous size. *Megaceros giganteus* grew to 4m (13ft) tall, with antlers of up to 3.5m (11½ft) across and 50kg (110lbs) in weight. It is believed that this impressive species originated in Siberia and slowly moved west to avoid the excesses of the weather. *Megaceros giganteus* first emerged during the Pleistocene Epoch and finally became extinct around 5700BC. The largest collection of giant deer skeletons can be seen in the Natural History Museum in Dublin.

Order:	Artiodactyla
Family:	Cervidae
Habitat:	Wooded grasslands
Distribution:	Europe and Western Asia
Time scale:	Pleistocene–Holocene

Merycoidodon sp.

If you were to visit the American plains 35 million years ago, you would probably see vast roaming herds of this short, stocky ruminant. (A ruminant is any animal which 'chews the cud'. They can regurgitate their food from one of four special compartments in their stomach so that they can re-chew it. In this way, they are able to extract as much nutrition as possible from even the poorest foods.) *Merycoidodon* was a distant relation of the modern-day camel, although it looked more like a short-legged deer. The 'type' species, whose features and characteristics became the yardstick by which other members of the genus were identified, is *Merycoidodon culbertsonii*. This was discovered in 1848 by fur trader Alexander Culbertson (1809–1879).

Order:	Artiodactyla
Family :	Merycoidodontidae
Habitat:	Plains & forests
Distribution:	North America
Time scale:	Mid-Oligocene

Poebrotherium wilsonii

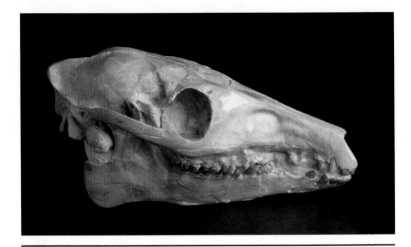

When the American palaeontologist Joseph Leidy (1823–1891) first saw an example of *Poebrotherium* in 1848, he was baffled. That's not so surprising when we consider that camels generally conjure up images of wind-swept desert sand dunes. Yet *Poebrotherium* – the ancestor of today's modern camels – made their homes in the woodlands of North America. This distinctive-looking species was less than 1m (3¼ft) long, with a narrow snout and long neck – a little like a Latin American Llama. Build for speed rather than endurance, *Poebrotherium* had long slender legs that ended in a set of splayed toes. In common with many herbivores, its teeth extended forwards from the jaw, allowing it to pull up vegetation easily as it grazed.

Order:	Artiodactyla
Family:	Camelidae
Habitat:	Plains and forests
Distribution:	South Dakota, USA
Time scale:	Oligocene

Cetacean

The terrestrial ancestors of whales started making their way back into the sea early in the Tertiary, shortly after the large marine reptiles of the Mesozoic had gone extinct 65 million years ago, leaving their marine niches vacant. Cetaceans quickly spread and diversified, and one lineage ultimately evolved into the highly successful, and often gigantic, baleen whales, which filter their food using a method unique among mammals: long hanging slabs of keratin line the upper jaws and sieve the whale's food on their frayed edges. The fossil pictured here shows a string of whale vertebrae and partial ribs. This is clearly not how this fossil was found – the bones were apparently placed together like this to reconstruct a disarticulated, incomplete, and broken skeleton from the recovered bones.

Class:	Mammalia
Order:	Cetacea
Habitat:	Marine
Distribution:	Global
Time scale:	Eocene–Recent

Squalodon sp. (Jaws)

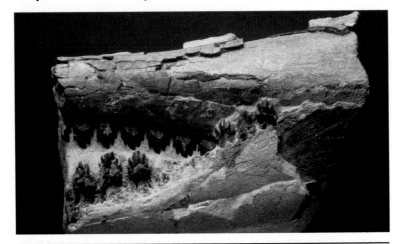

This primitive whale was common in the oceans of the world 18 million years ago. Now extinct, this genus belongs to a group of whales who get their name from the shape of their teeth. In Greek, 'squalus' means shark and 'don', tooth. So squalodonts are shark-toothed whales. The fossil shown above comes from a layer of sandstone rocks in the Southern Alps, which dates back to the Miocene Epoch. As can be seen, these formidable hunters had a particularly elongated skull (measuring around 80cm (31½in) on average) and a pointed snout. Inside the snout are rows of fearsome-looking teeth, designed for tearing apart prey. These have a serrated (jagged) edge, like those of a shark.

Order:	Cetacea
Family:	Squalodintidae
Habitat:	Marine
Distribution:	Global
Time scale:	Oligocene–Miocene

Cetacean (earbone)

Complete skeletons are relatively rare as fossils. This is especially true of very large animals that are subject to scavenging, decay and weathering on a very large scale. It is much more common to find isolated elements of skeletons and, not surprisingly, the most durable parts have the best chance of being preserved. Such is the case for extremely dense and compact whale earbones, represented by two bones called the tympanic and the periotic bullae. In some marine deposits, like the Tertiary sediments of South Carolina in the Unite States, where this tympanic bulla comes from, they are very common fossils. The white tube adhering to this specimen is a modern worm tube that grew on the fossil after it was exposed and was once again lying on the sea floor.

Order:	Cetacea
Family:	n/a
Habitat:	Marine
Distribution:	Global
Time scale:	Eocene–Recent

Desmostylus hesperus (tooth)

Desmostylians, of which *Desmostylus* is the classic example, was a small group of semi-aquatic mammals that existed for a short time on the shores of the North Pacific. They have been likened to marine hippos, in reference to their amphibious habit, proportions, and quadrupedal stance, but were more closely related to elephants. Skeletons of desmostylians are very rare but deposits like the Tremblor Formation, of Kern Co., California, have produced large numbers of teeth, such as this one, showing desmostylians were a conspicuous part of the Miocene coasts in the area. Their chewing teeth were unlike any other mammal's – each consisted of a group of columnar elements that were ground down perpendicular to their long axes by the coastal plants, and possibly invertebrates, the animal ate.

Order:	Desmostylia
Family:	Desmostylidae
Habitat:	Coastal regions
Distribution:	Pacific coastline of USA & Japan
Time scale:	Miocene

Moeritherium sp.

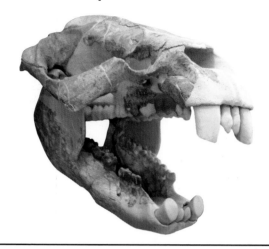

*M*oeritherium means 'Moeris Beast', after the oasis in Fayum, Egypt, where this primitive mammal's fossils were first discovered in 1904. Although related to modern-day African and Indian elephants, *Moeritherium* looked more like a long-snouted pig – measuring around 0.7m (2¼ft) tall, with short, stocky legs, a 3m (10ft) long body and broad hoofed feet. *Moeritherium* made its home in marsh and swamp lands and, as can be seen from this reconstructed fossilized skull, its eyes and nostrils are positioned high on its head. This suggests that it spent much of its time semi-submerged. Although *Moeritherium* did not have a trunk, its proboscis was long and flexible enough to be used like a trunk to uproot aquatic plants and grasp floating vegetation as it grazed.

Order:	Proboscidea
Family:	Moeritheriidae
Habitat:	Coastal swamps & marsh lands
Distribution:	North Africa
Time scale:	Late Eocene–Early Oligocene

Mammuthus sp.

The North American mammoth is, when reconstructed, one of the most easily recognisable of all prehistoric animals: it was a primitive relative of the modern elephant. *Mammuthus* stood about 2.5m (8ft) tall at the shoulder and had thickly enamelled cheek teeth with which to crush its food. The remains of mammoths can still be seen in situ at the Mammoth Site in the United States where a great find was made at Hot Springs, South Dakota in 1974. First revealed were three woolly mammoths from a 26,000-year old sinkhole that soon yielded other Ice Age animals, such as the llama, giant bear, wolf, coyote and thousands of mollusc shells.

Class:	Mammal
Order:	Proboscidea
Habitat:	Terrestrial
Distribution:	Northern Hemisphere
Time scale:	Pleistocene (The Mammoth Site)

Mammuthus primigenius (Left eye skin)

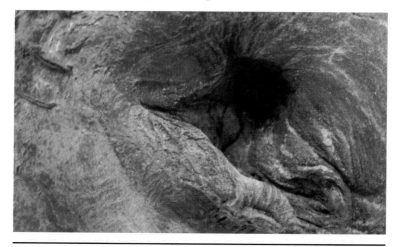

When most people think of fossils, they think of the hard part remains of organisms that have turned to, or are preserved in stone. Although the vast majority of fossils could be described this way, there are extraordinary examples which preserve soft parts, and without any mineralization at all. Some Pleistocene animals are preserved in frozen mud, or permafrost (as opposed to blocks of ice) and if buried before decay had set in, can preserve organs, skin, hair, blood and even DNA. This specimen is just the left eyelids of a partial baby mammoth found in Alaska in 1948. Named 'Effie,' in honour of the workers of the Fairbanks Expedition (FE) branch of the United States Smelting, Refining, and Mining Corporation who discovered it, it shows its face, including the trunk, and one leg.

Order:	Proboscidea
Family:	Elephantidae
Habitat:	Steppe grasslands
Distribution:	Northern terrestrial latitudes
Time scale:	Pleistocene

Mammal (coprolites)

The first coprolites were recognized as such by William Buckland in the 1820s. Identifying coprolites requires observation of surrounding context and knowledge of modern animal droppings. Several features of these coprolites, and others of their kind from the White River Badlands of South Dakota, strongly suggest that they were made by carnivorous mammals. First, they are found in sediments that were once terrestrial and contain abundant fossils of mammals. Second, their shape is very similar to modern carnivorous mammal faeces. And third, they contain lots of phosphate – which is to be expected from an animal ingesting lots of bone – as well as containing bones themselves. Some of the first coprolites identified by Buckland include ones likely made by carnivorous animals.

Class:	Mammalia
Order:	Carnivora
Habitat:	Various
Distribution:	Global
Time scale:	Triassic–Recent (mammals); Oligocene (these coprolites)

Palaeochiropteryx sp.

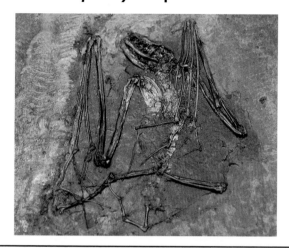

Messel, Germany, boasts a fantastic fossil site of Eocene age, whose oil shales wonderfully preserve a diverse fauna that includes such rarities as birds with feathers, whole snake skeletons, mammals with fur and gut contents, and beetles that retain their iridescence. But the rocks degrade quickly in air and need very special conservation to save the contained fossils. The layered rock is prepared so that the flattened fossil is exposed on a flat surface. This is then embedded in a usually clear plastic or epoxy, after which the still rock-enclosed portion is freed of rock. In the end, one is left with a transfer prepared specimen – a fossil kept in its original position but relocated to another, artificial, matrix – like this complete Messel bat.

Order:	Chiroptera
Family:	Palaeochiropterygidae
Habitat:	n/a
Distribution:	Global (bats)
Time scale:	Eocene–Recent (bats)

Oreopithecus bambolii

When our ancestors were still swinging through the trees over eight million years ago, *Oreopithecus bambolii* had already taken its first steps on two feet. Although it was probably a slow walker, its S-shaped spine and long upper leg bones strongly suggest that *Oreopithecus* was a true biped – walking on two legs much of the time rather than just occasionally, like chimpanzees. This makes our ancestors not the first, nor even the earliest, primate to make that vital evolutionary leap. Remains of 50 of these ancient 'swamp apes' have so far been found in Tuscany, which was once an island. It is believed that *Oreopithecus* survived and thrived there for around two million years. When a land bridge joined the island to the mainland, this little ape quickly fell prey to predators entering their habitat.

Order:	Primates
Family:	Oreopithecidae
Habitat:	Swamps
Distribution:	Italy and East Africa
Time scale:	Miocene

Australopithecus afarensis (model)

Until the discovery of *Australopithecus anamensis* in 1995, this small, agile hominid was our earliest-known ancestor. Hominids are the group to which humans belong. This model shows how *Australopithecus* may have looked. In some ways, it was similar to a modern day chimpanzee, with a compact body, long arms and a small skull. In common with many apes, there is also evidence that males of the species were significantly larger than females. The most important difference between *A. afarensis* and modern apes, however, is that they walked upright. In fact, their feet and pelvis are almost identical to those of humans. So far, only 300 examples have been discovered, the most famous being an almost complete skeleton, nicknamed 'Lucy', found in Hadar, Ethiopia in 1974.

Order:	Primates
Family:	Hominidae
Habitat:	Lightly forested areas, bush & grasslands
Distribution:	North East and Southern Africa
Time scale:	Pliocene

Homo neanderthalensis (stone tools)

Early humans were tool makers, and evidence of their skills can be found throughout Europe, the Near East and North Africa. The earliest and simplest tools were made by hitting a suitable stone with a harder hammerstone until a chipped edge became sharp. However, by the time of *Homo erectus*, soft hammers made from wood or bone were being used to shape the stone into more complex hand axes and knives. Pictured here are Mousterian tools, which date from around 130,000 years ago – the time of Neanderthal man. Made mainly of flint, they get their name from Le Moustier, in France, where many examples have been discovered. Most Mousterian tools are made using the 'Levallois Technique' which leaves one side flat and the other slightly domed.

Order:	Primates
Family:	Hominidae
Habitat:	Various
Distribution:	Un-glaciated Europe, the Near East and North Africa
Time scale:	Pleistocene

Glossary

Ammonoid	One of a group of extinct, coil-shelled cephalopods made up of three subgroups: ceratites, goniatites, and ammonites
Brackish	Water that is part way between fresh and salt
Commensalism:	Any interaction between two organisms in which one benefits and the other is neither harmed nor helped
Cyanobacteria	Photosynthetic bacteria (previously known as blue-green algae)
Deciduous	Used to describe plants that lose all of their leaves for part of the year
Formation	A formally named unit of rock of a single type that can be mapped
Gondwana	An ancient continent of the southern hemisphere composed of the land masses of South America, Africa, India, Australia and Antarctica
Gymnosperm	Seed bearing plant with 'naked seeds' e.g conifers, ginkgos and cycads
Laurasia	An ancient continent of the northern hemisphere composed of the land masses of North America Europe and Asia (minus India)
Matrix	The enclosing sediment in which a fossil is found
mya	Abbreviation for 'million years ago' pronounced 'm-y-a'
Nektonic	Freely swimming
Nomenclature	Naming
Organism	Any living thing
Paleobotany	The study of fossil plants
Pangaea	An ancient supercontinent composed of all existing landmasses that came into existence in the Late Paleozoic and broke up in the Mesozoic
Parasitism	Interaction between two organisms where one benefits and one is harmed
Permafrost	Ground that is frozen throughout the year
Pinnate	In botany, used to describe leaves that are shaped like a feather
Polyp	A coral individual, as opposed to a colony
Proboscis	An elongated appendage from the head of an animal
Sessile	In zoology, unable to move and, generally, attached
Silicify	To become converted into or impregnated with silica, or silicon dioxide
Spat	Young oysters
Steppe	A dry plain with short grass and devoid of trees
Stratigraphy	Rock layers of the study of these layers
Symbiosis	Any interaction between two organisms in which both benefit
Tundra	A geographical area where tree growth is hindered by low temperatures and short growing seasons

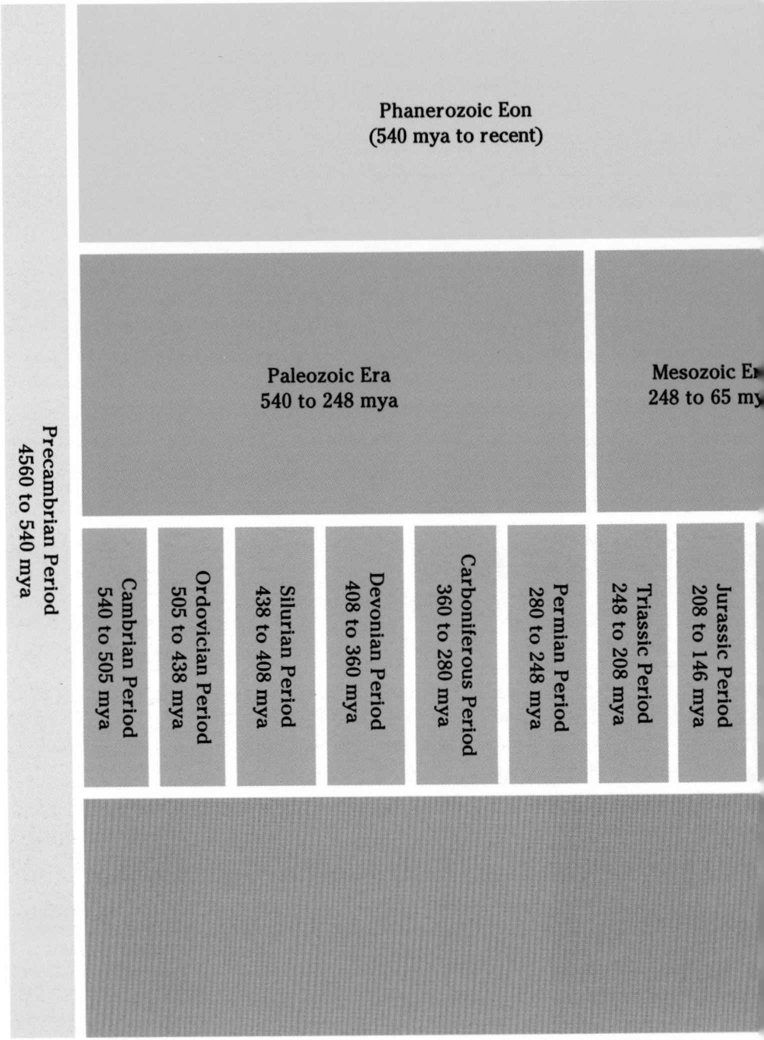

Phanerozoic Eon
(540 mya to recent)

Paleozoic Era
540 to 248 mya

Mesozoic Era
248 to 65 mya

Precambrian Period
4560 to 540 mya

Cambrian Period
540 to 505 mya

Ordovician Period
505 to 438 mya

Silurian Period
438 to 408 mya

Devonian Period
408 to 360 mya

Carboniferous Period
360 to 280 mya

Permian Period
280 to 248 mya

Triassic Period
248 to 208 mya

Jurassic Period
208 to 146 mya

mya = million years ago

Cenozoic Era
(65 mya to recent)

Tertiary Period
(65 to 1.8 mya)

Quaternary Period
(1.8 mya to recent)

Palaeocene
Epoch
65-54 mya

Eocene Epoch
54-38 mya

Oligocene Epoch
38-24 mya

Miocene Epoch
24-5 mya

Pliocene Epoch
5-1.8 mya

Pleistocene Epoch
1.8-.011 mya

Holocene Epoch
0.11 mya
to recent

Index

INDEX

The editor wishes to thank the following people for their help and support:
John Alroy, Robert Asher, Fiona Brady, Dan Brumbaugh, Sandra J. Carlson, Charles Catton, Mick Ellison, Mike Everhart, Norberto P. Giannini, David Grimaldi, Ian Harrison, Ralph Johnson, Neil Landman, Alexander, Lukeneder, Ross Macphee, John G. Maisey, Barbara Mathe, Duncan McIlroy, Mary & Walter Mehling, Jim Mills, Nancy Simmons, Alan Turner, Alfred Uchman, Sarah Uttridge.